98
S0-BUH-674

PETRIFIED WOOD

The World of Fossilized Wood, Cones, Ferns, and Cycads

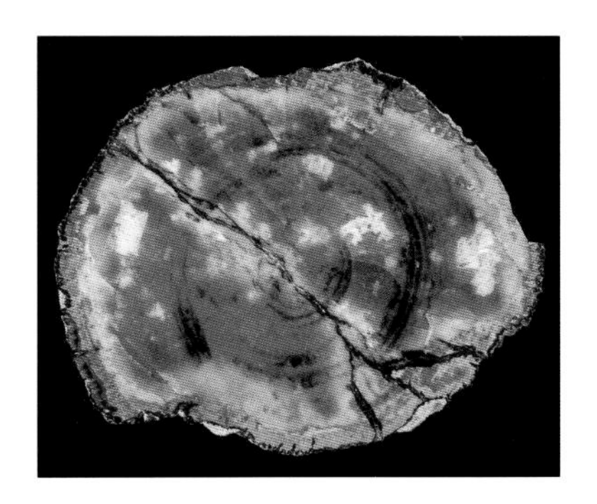

Frank J. Daniels

F. Daniels #230

EDITORS

Brooks B. Britt, Ph.D. Paleontology; Curator of Paleontology, Museum of Western Colorado

Richard D. Dayvault, M.S. Geology

WESTERN COLORADO PUBLISHING COMPANY

First Edition

Any individual or institution having information or specimens that may be appropriate for a subsequent volume, or anyone wishing to order additional copies of this book, may contact the publisher at the address below:

WESTERN COLORADO PUBLISHING COMPANY
2024 Freedom Court
Grand Junction, Colorado 81503-9522
(970) 242-5255
e-mail: cycadwood@aol.com

Printed by Pyramid Printing, Inc., Grand Junction, Colorado
Design by Frank J. Daniels and Kitty Anderson Nicholason

♾ The 100% acid-free paper used in this publication meets the minimum requirements of the American National Standard for Information Sciences-Permanence of Paper for Printed Library Materials.

Cataloging-in-Publication Data prepared by Western Colorado Publishing Company

Daniels, Frank J. 1949-
Petrified wood : the world of fossilized wood, cones, ferns, and cycads / Frank J. Daniels

Includes bibliographical references, glossary, and index.
ISBN 0-9662938-0-0
1. Trees, Fossil. 2. Plants, Fossil. 3. Cycads, Fossil.
4. Petrification. 5. Paleobotany. I. Title.
QE991 .D36 1998
561.21 dc-20 561.16 dc-21
Library of Congress Card Catalog Number: 98-90104

Published and printed in the United States of America

Preface

"Rock gives reality
to the otherwise abstract notion
of transhuman time."

Edward Abbey

There are a number of ways to gain perspective on one's place in the universe. For me, petrified wood is one. While a beautiful, well-silicified, gem quality, colorful branch of petrified wood is now a rock, it once was part of a tree—a tree that may have been growing in a distant forest over 200 million years ago. Some of these trees grew when the continents of the earth were joined into one. It is difficult to imagine the events that allowed these petrifications to occur and the forces that later allowed them to be unearthed. Each specimen has a story of its own.

Professor George R. Wieland of Yale University and the Carnegie Institution of Washington, whose works are referenced throughout this volume, was a pioneer in paleobotany. His interest in petrified wood, cones, ferns, and cycads resulted in several of the earliest and most important treatments of the subject. The continuing relevance of his works and their historical reference illustrate how the study of these materials can lead one to a heightened appreciation and knowledge of fossilized flora. A comparison of the plates from Professor Wieland's works and those in this volume underscores the advances that have occurred in photography and photographic reproduction.

The purpose of this book is to display the beautiful colors and textures of petrified wood, cones, ferns, and cycads, and to give the reader a taste of the many disciplines for further study. Depicted in the photographs are some of the finest cut and polished specimens in existence.

Frank J. Daniels
Grand Junction, Colorado
June 1998

Wolverine Petrified Wood Natural Area in Grand Staircase – Escalante National Monument, Utah.

Pseudotsuga menziesii **[Douglas fir] in Bryce Canyon National Park, Utah.**

Petrified Forest National Park. Established as a national monument by President Theodore Roosevelt in 1906, the area was enlarged to become a national park in 1962.

Acknowledgements

Specimens for this book are from a variety of sources. A number of individuals generously contributed their time and the use of their most beautiful pieces. Photograph captions provide the following information about the specimen(s) when known with a reasonable degree of certainty: the area from which the specimen was collected, including the state, country, and locality; the geologic formation and corresponding time unit; an identification of the specimen; the specimen owner; and the specimen size as measured across the widest dimension shown to the nearest half centimeter.

I am especially grateful to the following major contributors. Each owns an important collection or inventory of rare and spectacular petrified specimens which he or she generously shared with the author. The knowledge imparted to me by these individuals was invaluable to the preparation of this work.

E.P. Akin
Shreveport, Louisiana
Louisiana petrified *Palmoxylon*

John Bennett
Perth, Western Australia
Australian petrified wood

Bill Branson
Helper, Utah
Utah petrified *Cycadeoidea*

Jim Gray
Holbrook, Arizona
Arizona petrified wood

Steve Hatch
Hanksville, Utah
Utah petrified *Cycadeoidea* and cones

Vince Jones
The Vince Jones Collection
Grand Junction, Colorado
Utah and Wyoming petrified wood

Nyla Kladder
Grand Junction, Colorado
Yellow Cat, Utah carnelian "redwood"

Barry Lark
Queensland, Australia
Australian petrified wood

Dan Rigel
Grants Pass, Oregon
Northwest United States petrified wood &
Australian petrified wood and ferns

Tom Robertson
Salem, Oregon
Oregon petrified wood

Bill Rose
Vantage, Washington
Washington petrified wood

J.B. Sanchez
Cortez, Colorado
Colorado petrified *Hermanophyton*

Gratitude is extended to the following individuals for the use of their materials and for the information and support provided:

Phil and Joni Andrist, Bandon, Oregon; Bob and Dan Beck, Mount Hood, Oregon; Dominic Cataldo, Kennewick, Washington; Ana and Luis De Los Santos, Buenos Aires, Argentina and Fontana, California; Ulrich and Dörte Dernbach, Heppenheim, Germany; Ken and Jeanie Emmons, Redding, California; Ronald P. Geitgey of the Oregon Department of Geology and Mineral Industries; Ron Justman, Holbrook, Arizona; Phil and Emma Johnson, Sparks, Nevada; Al Keller, Olathe, Colorado; Dean Lowdermilk, Grand Junction, Colorado; Adam Luna, Holbrook, Arizona; Ray Lyman, Blanding, Utah; Henry and Marie Macom, Grand Junction, Colorado; Jim and Julie Mitchell, Deming, New Mexico; Dennis and Mary Murphy, Tigard, Oregon; Arnold Nottingham, Grand Junction, Colorado; Glen Pryor, Grand Junction, Colorado; Joan Schubarth, Grand Junction, Colorado; Chuck Scott, St. Johns, Arizona; and Ernie Shirley, Hanksville, Utah.

Australian specimens belonging to John Bennett and Barry Lark were photographed in Australia by an anonymous photographer and by John Milne, respectively. All other photographs were taken by the author. No color enhancing filters were used in any of these photographs. Most shots were taken in natural sunlight.

Sincere gratitude is also extended to the staff of the Mesa County Public Library in Grand Junction, Colorado, for their assistance in obtaining research materials; to the Carnegie Institution of Washington for permission to use portions of the works of G.R. Wieland; to all members of the Grand Junction Gem and Mineral Club; and most specially to my family, Martelle, April, Hillary, Prudence, and Olivia, for all they endured while I worked on this book.

Identification of fossil woods is an exacting scientific process, requiring materials to be cut into cross, radial, and tangential thin sections ground to a thickness of one or two cells and examined under a high power microscope. Identification is based largely on comparison with living woods. A significant number of the photographed specimens are not identified. The best petrified materials for polishing are often the poorest for identification. Wood cells that have undergone significant alterations due to chemical or mechanical geologic processes may be too altered or compressed to permit identification. Fossil casts and completely replaced specimens, such as the carnelian "redwood" of Yellow Cat, Utah, are impossible to identify because there are no cells present in any form.

Contents

Triassic and Jurassic formations meet at the Chinle-Wingate contact seen at the base of these sheer sandstone cliffs in Colorado National Monument.

Petrified logs in Petrified Forest State Park, Escalante, Utah. Many large and colorful logs from the Jurassic age Morrison Formation are here exposed.

Introduction

"These depths are now so high that they have become hills, or high mountains, and the rivers that wear away the sides of those mountains lay bare the layers of these fossils."

Leonardo Da Vinci, 1452-1519

Gaze into the polished window of a petrified specimen and encounter a mysterious and beautiful ancient world, now forever gone. What strange and magnificent creatures ate the leaves that once grew from this plant? How appeared the sky that held our same sun to make this growth possible? Ponder the state of the earth as millions of years passed while the petrifying branches shifted, as if magically, from vegetable to mineral, and consider the mechanisms involved in forming the crystals and colors that now compose what was then simply a tree.

Reflected on by Leonardo Da Vinci and others, these questions are still the subject of serious inquiry. Such inquiries direct one down the paths of many disciplines: geology, mineralogy, paleobotany, paleoentomology, chemistry, and others.

***Araucarioxylon arizonicum* logs in Petrified Forest National Park, Arizona.**

Knot and lichen on a petrified log section in Petrified Forest State Park in Escalante, Utah.

Eroded cliffs within Capitol Reef National Park. The sheer Wingate Formation cliffs at the top and the dark red-brown Moenkopi Formation cliffs at the bottom sandwich Chinle Formation clays.

One of many; Petrified Forest National Park, Arizona.

Pinus ponderosa [Ponderosa pine].

The materials that are the subject matter of this book lived millions of years ago and traveled tens of thousands of miles to come before the lens of a camera. The polished specimens depicted here represent fossils that have become gems.

Of the billions of trees ever to have photosynthesized under the sun, only a minute fraction were placed in the circumstances to fossilize and only a small fraction of those survived. Many are incompletely silicified, fractured, or without color, leaving a barely calculable percentage in the category of those shown here. It is no wonder that understanding the processes of their creation leads to greater respect and admiration for these specimens. Which of the billions of trees living on the planet today will become such gems?

Wolverine Petrified Wood Natural Area in Grand Staircase–Escalante National Monument, Utah.

AMERICAN FOSSIL CYCADS

BY

G. R. WIELAND

PUBLISHED BY THE
CARNEGIE INSTITUTION OF WASHINGTON
AUGUST, 1906

Title page from *American Fossil Cycads,* written by Professor G.R. Wieland and published by the Carnegie Institution of Washington in 1906.

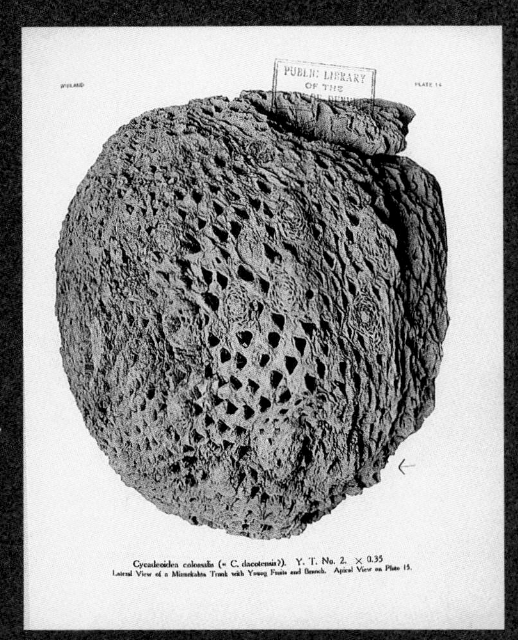

Plate of a Black Hills fossilized *Cycadeoidea, American Fossil Cycads, Volume II*, 1916.

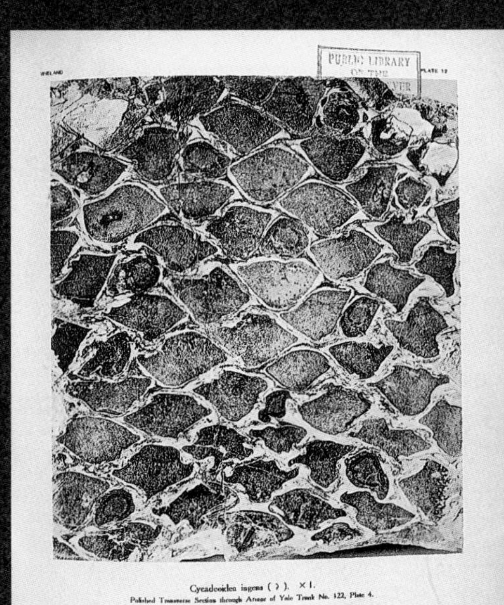

Plate of a fossilized *Cycadeoidea* section, *American Fossil Cycads, Volume II,* 1916.

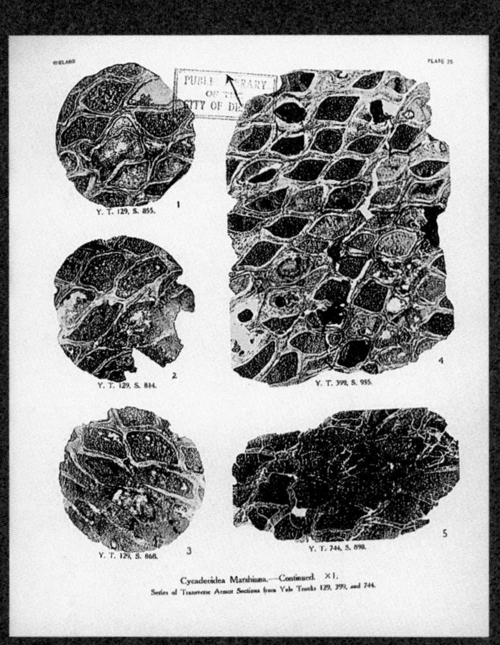

Plate of a set of *Cycadeoidea* sections, *American Fossil Cycads, Volume II*, 1916.

THE CERRO CUADRADO
PETRIFIED FOREST

By G. R. Wieland
Research Associate of Carnegie Institution of Washington

Published by Carnegie Institution of Washington
April 1935

Title page from *The Cerro Cuadrado Petrified Forest,* written by Professor G.R. Wieland and published by The Carnegie Institution of Washington in 1935.

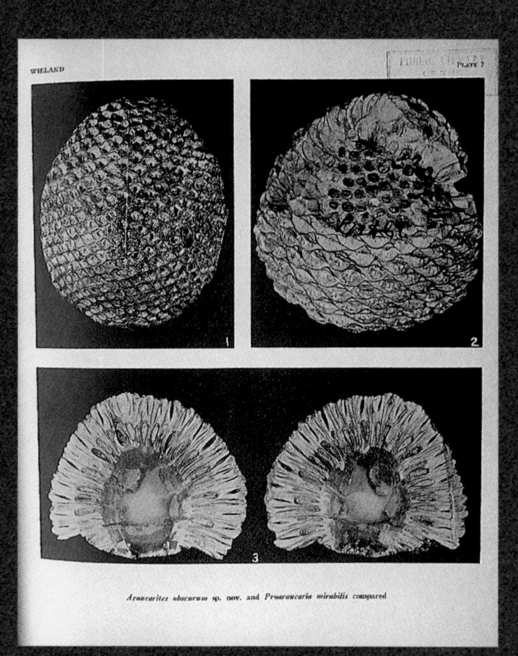

Plate of a set of fossil *Araucaria* cones from *The Cerro Cuadrado Petrified Forest,* 1935.

1 Paleobotany

"On the mountains and in all the valleys its branches will fall, and its boughs lie broken in all the watercourses of the land...."
Ezekiel 31:12

Paleontology is a science that developed late relative to other scientific disciplines. The study of paleobotany as a separate science began in Europe only within the past two hundred years. That is not to say that specimens of petrified wood and other materials were ignored by earlier civilizations. A fossilized cycad trunk was discovered in a sepulchral chamber in the ancient metropolis at Marzabotto, Italy, having been placed there by Etruscans more than 4,000 years ago, together with vases and other objects of superstitious reverence. Native Americans had various beliefs about the origin of the petrified logs in what is now Petrified Forest National Park in Arizona. Natives of the Paiute tribe held that these giant petrifications were spent arrow shafts and spears dispatched by the Thunder God Shinauav and his enemies during a great battle. Members of the Navajo tribe believed they were the bones of the great giant monster Yeitso. It is clear from the study of artifacts found in Native American ruins that petrified wood was used for a variety of purposes. Silicified petrified materials made excellent arrowheads and spearheads and were also used as building materials, game balls, and tools. These materials were doubtlessly used for similar purposes by aboriginal peoples worldwide. Leonardo Da Vinci was interested in petrified materials and used his knowledge of them

Arrowheads made from silicified wood by Native Americans in northern Arizona.

Petrified wood with fossilized tree fungus pockets.
Arizona, USA [St. Johns]
Mesozoic; Triassic
Araucarioxylon arizonicum **fm: Chinle**
Daniels 34 cm

Petrified wood with insect borings, probably made by the larval stage of the flat-headed borer.
Wyoming, USA [Eden Valley]
Cenozoic; Tertiary; Eocene
unidentified fm: Green River
Daniels 12 cm

A cross-section of petrified oak which appears to have petrified after the center of the log rotted away.
source unknown
Daniels 20 cm

Petrified wood section showing evidence of insect damage.
Wyoming, USA [Eden Valley]
Cenozoic; Tertiary; Eocene
unidentified fm: Green River
Daniels 9.5 cm

in developing theories regarding the creation of the earth. Charles Darwin studied localities containing petrified materials; after viewing a formation containing petrified wood in Argentina in 1835, he penned:

> *Nor had those antagonistic forces been dormant, which are always at work wearing down the surface of the land: the great piles of strata had been intersected by many wide valleys and the trees, now changed into silex, were exposed projecting from the volcanic soil, now changed into rock, whence formerly, in a green and budding state, they had raised their lofty heads.*

Plate of a collection of fossilized Black Hills *Cycadeoidea, American Fossil Cycads, Volume II,* 1916.

John Muir, an American naturalist more noted for his efforts to preserve the living redwood forests of California, spent a considerable amount of time studying petrifications in what is now Petrified Forest National Park. Muir collected a variety of fossil specimens and returned them to Berkeley for further study. In 1879, General William Tecumseh Sherman, of Civil War fame, collected petrified wood from the area and returned several specimens to the Smithsonian Museum. Professor George R. Wieland of Yale University was an early researcher in the study of petrified plant materials who, near the turn of the century, turned his attention to the petrified cycads of the Black Hills of South Dakota and Wyoming. At about this same time, an American interest in preserving some of these petrified materials began to take hold. In 1906, President Theodore Roosevelt, using powers granted to him by the Antiquities Act, set aside as the nation's second national monument an area that constitutes a large portion of what in 1962 became Petrified Forest National Park. America's first national park, Yellowstone, established in 1872, also contains large and interesting petrified forests, and America's newest national monument, Grand Staircase–Escalante National Monument, established in 1996, reveals many petrified wood deposits, Triassic to Cretaceous in age. Among these deposits is Wolverine Petrified Wood Natural Area, a large and scenic expanse replete with many striking petrified logs.

Of all the plants that have grown upon the face of this planet, only a small fraction have achieved the distinction of becoming fossilized. As Professor Wieland noted in his 1906 work, *American Fossil Cycads*, "The conditions requisite for silicification must have seldom occurred in geologic time." Concerning the process of petrification, Professor Wieland wrote:

> *The mineralization of entire plants as fossils is conditional, firstly, upon the tissue systems present, as connected with particular states of growth, prefoliations, preflorations, or fructification. The secondary factors of control in the process are mainly: (1) the relative abundance and kind of mineralizing materials; (2) temperature; (3) the presence or absence of secondary reagents, such as iron, capable of replacing plant tissues and preserving their microscopic structures in finely differentiated form, but not necessary to the process of silicification or calcification, as the case may be; (4) the duration and rapidity of chemical activity; (5) the nature of the embedding rock material. These are the principal elements determining the clearness with which structural details are preserved and differentiated, the final results being in addition dependent upon the freedom from maceration or decay of the original plants at the time of their fossilization, as well*

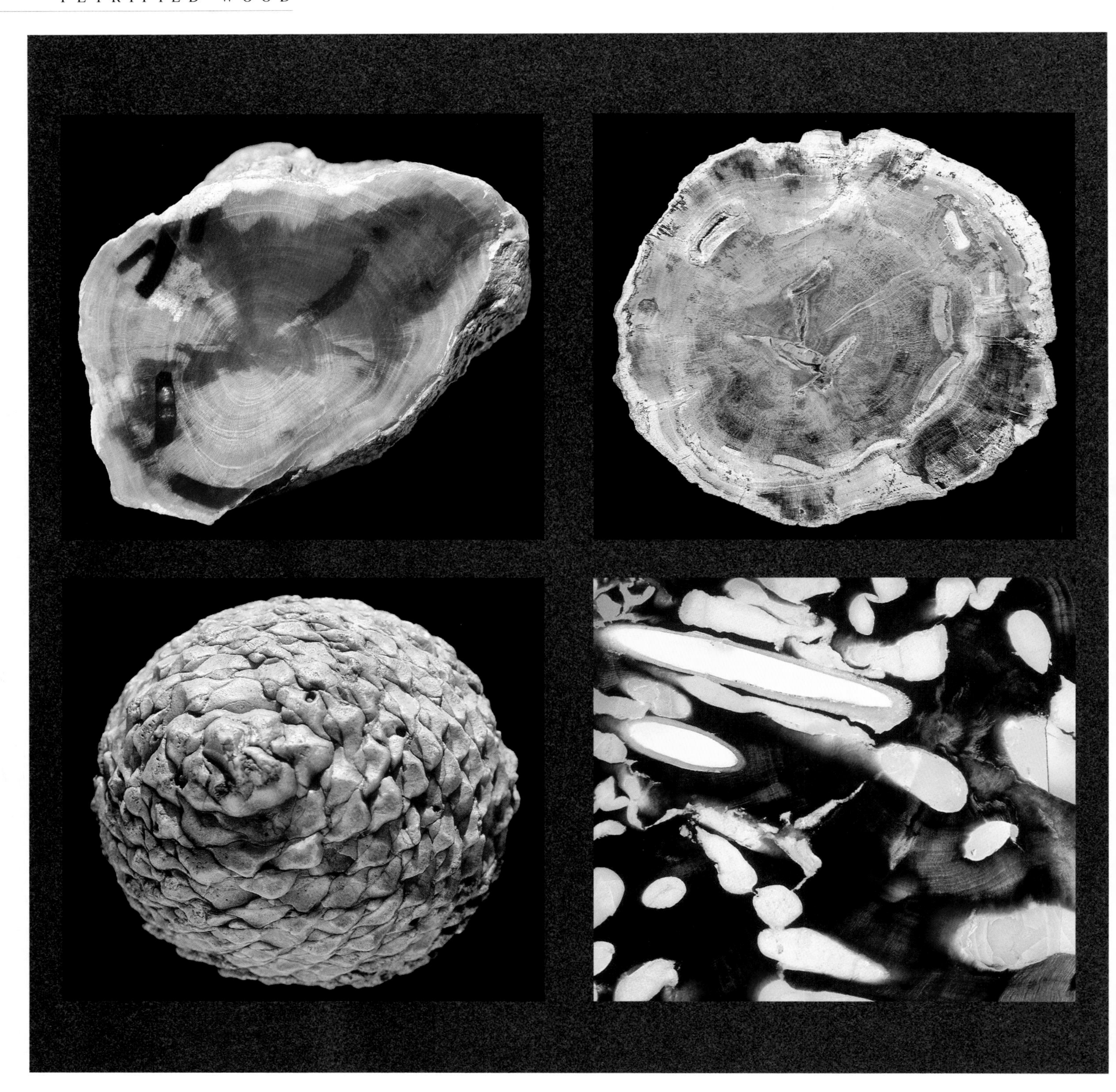

Petrified wood in cross-section showing what appears to be a petrified larva in one of several chalcedonized bore tunnels.
Nevada, USA
Cenozoic; Tertiary; Miocene
***Acer* [maple] fm: Trout Creek**
Rigel 7 cm

Cross-section of petrified wood showing insect borings.
Zimbabwe, Africa
Mesozoic; Triassic
Araucaria
Daniels 11 cm

Petrified *Araucaria* cone with insect damage.
Patagonia, Argentina
Mesozoic; Jurassic
Araucaria mirabilis
Daniels 6 cm

Cross-section of petrified wood showing silicified bivalve mollusk bore tunnels.
Australia
Mesozoic; Jurassic
conifer [shipworm wood]
Daniels

as from subsequent chemical changes or compression in the containing beds after the early process of preservation is completed.

While Professor Wieland's explanation of petrification was written at the very beginning of the century, it is not today subject to substantial challenge. The petrification process proceeds as an infiltration of cell structures and interstices by mineral-laden solutions. There must be a source of sedimentary material in a soluble or particulate form. An early petrification will retain a considerable amount of the original material of the plant, albeit now encased in mineral. As time proceeds, increasingly more of the original matter is replaced by the invading minerals, often resulting in what Professor Wieland referred to as a petrification that has become "too over-chalcedonized to reach distinctions."

Transitional forest on Boulder Mountain in Dixie National Forest showing a mix of gymnosperm conifers and angiosperm aspens. Eventually the aspen will give way to the conifers.

One obvious reason for the relative rarity of petrified materials, given the enormous abundance of sedimentary rocks on the earth, is that, in general, plant materials decay rather than petrify. For a tree to become petrified, it must be in an anaerobic environment to prevent decay, that is, it somehow must be buried in an oxygen free environment, possibly in the bend of a silt-laden river, in the bottom of a lake, or in what appears to be the most frequent manner, in volcanic deposits. An examination of the vast majority of the petrified forests of the world reveals a relationship with volcanic action. This can be by means of massive ash falls in the environment of petrification, as appears to have been the case in northern Arizona in the Chinle Formation; by the actual burial of a forest itself in volcanic ash, as appears to have occurred in much of the petrified forest in Patagonia, Argentina; or by the burial of a forest in a lava flow or its associated mud flows, as probably occurred in the Miocene and Eocene petrified forests in Oregon, Washington, and Wyoming. Professor Wieland wrote: "Silicification is a process admittedly demanding a fair freedom from decay of any kind either before or during replacement. It requires structure which is at least initially intact. It is also a process which appears to involve extremely dilute orthosilicic acid solutions, and may hence go on with an excessive, almost geologic slowness."

How much time is required for any particular organic material to petrify? Clear evidence of petrification can be seen in mine timbers from around the world, from 2,000-year-old Mediterranean copper mine timbers that appear, on the surface, to have completely turned to copper, to younger timbers in silver mines in Mexico and copper mines in Montana in the United States. These materials demonstrate that the process of petrification can begin in a relatively short period of time. Of course, these materials are not completely petrified and are not petrified in the same manner as are gem quality petrified woods. At the other end of the spectrum are some 100-million-year-old trees preserved in Cretaceous oil sands in northern Alberta, Canada. These trees are preserved in their near-original condition in a thick, black oil. The cell walls appear to be perfectly preserved, but they are not petrified. The trees can be cut with an ordinary carpenter's saw.

An interesting example of the formation of a petrified forest is that found in Yellowstone National Park in Wyoming. This forest was formed during the Eocene Epoch, some 45 to 50 million years ago. Petrified materials of more than eighty plant species have been discovered, the most common trees being walnut, sycamore, oak,

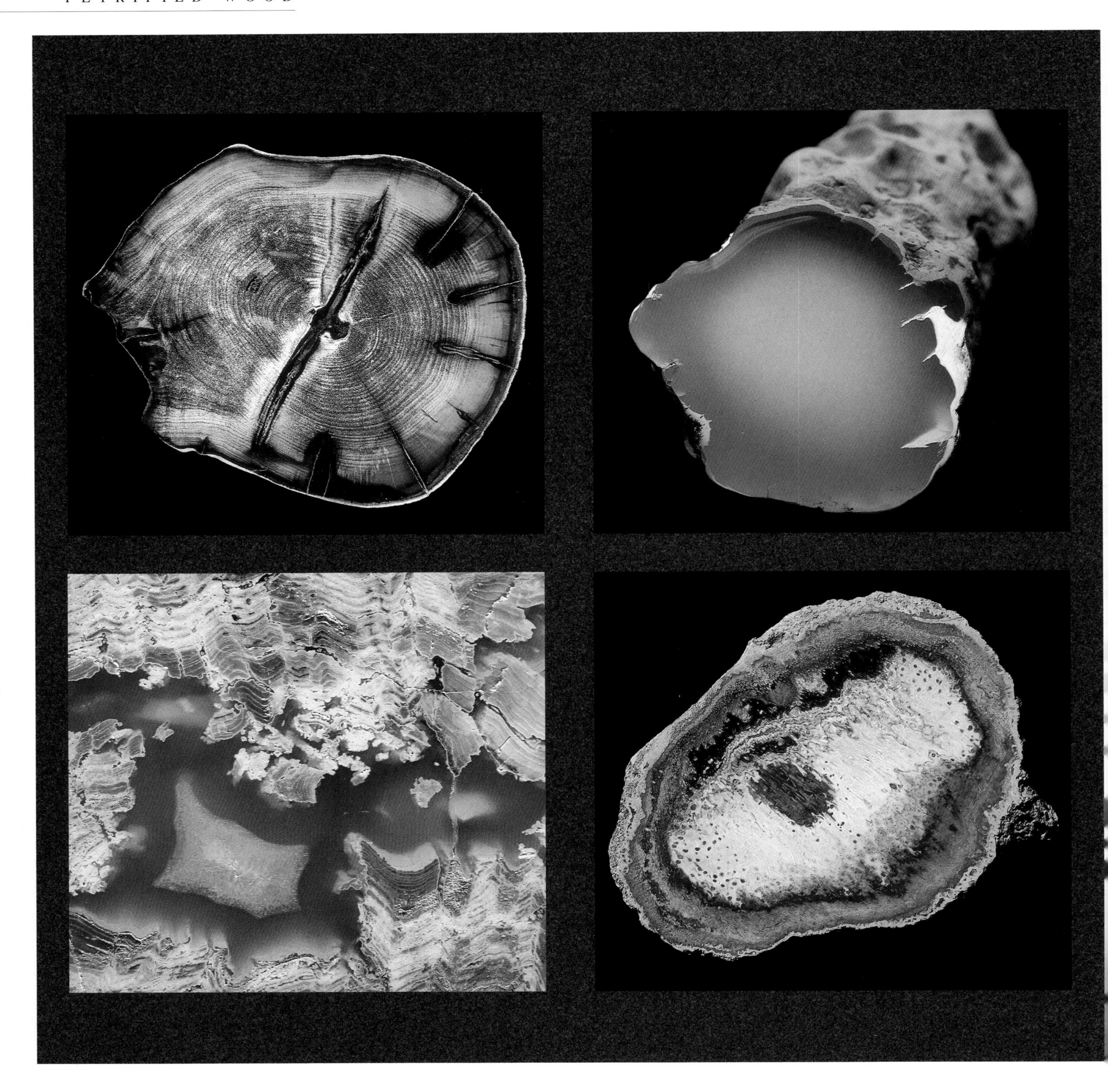

Petrified wood in cross-section showing detail of annual rings and cell structure.
Oregon, USA [Eagle's Nest]
Cenozoic; Tertiary
***Betula* [birch]**
Rigel 16 cm

Polished end of a cast.
Oregon, USA [Prineville]
Cenozoic; Tertiary
Daniels 4 cm

Petrified wood cross-section showing brecciated sections of permineralized wood "floating" in chalcedony.
Oregon, USA [Swartz Canyon]
Cenozoic; Tertiary
Daniels

Texas, USA
Cenozoic; Tertiary; Oligocene
***Palmoxylon* fm: Catahoula**
Daniels 35.5 cm

Polished specimen of copal, a fossilized tree resin, with imprint of the hollow of a tree.
Columbia, South America
Cenozoic
copal
Daniels 6 pounds

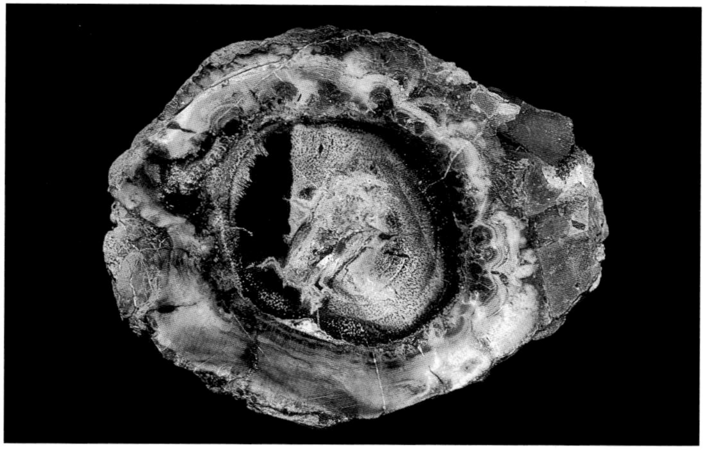

Petrified wood encased in petrified algae.
Oregon, USA [McDermitt]
Cenozoic; Tertiary; Miocene
spruce in algae
Rigel 8.5 cm

magnolia, redwood, chestnut, maple, persimmon, and dogwood. This petrified forest is actually a sequence of petrified forests at least twenty-seven forests deep. Apparently a forest grew for several hundred years to maturity and was buried in ash and mud flows. Mineral rich waters percolated through the layers to aid the petrification process. Each subsequent volcanic eruption buried the trees to a depth of 3 to 15 feet. The forest covers more than 24.5 square miles and contains over 1,000 vertical feet of volcanic material embedded with petrified wood, the stumps generally being preserved in a growth position.

As opposed to the upright standing trunks in the Yellowstone petrified forest, the trunks in the older, Triassic age petrified forests of Arizona are typically well worn, without limbs, bark, or roots, and are not generally found in conjunction with their seeds or cones. Apparently they were transported and deposited by flooding rivers during cataclysmic volcanic events that caused large amounts of volcanic ash to be included in the formation. The petrified forests of northern Arizona from the Triassic Period and of northern Wyoming from the Eocene Epoch of the Tertiary Period both were buried in one manner or another and exposed to large amounts of volcanic materials. Waters passing through overlying volcanic ash laden with silica preserved the wood.

In the Triassic age petrified forests in northern Arizona, the majority of the fossilized trees are *Araucarioxylon arizonicum*, *Woodworthia*, and *Schilderia*. *Woodworthia* and *Schilderia* specimens are only found in the upper Chinle rocks. *Woodworthia* specimens are generally smaller than *Araucarioxylon*, and *Schilderia* are smaller than *Woodworthia*. Fossilized trees similar to those in the park are also found in corresponding formations in South America and Africa, lending support to the theory that during the Triassic Period all continents were joined in the giant supercontinent Pangaea. A specimen of Triassic petrified wood from Argentina can be difficult to distinguish from one from Arizona. Araucarian logs both in the Arizona forest and in Patagonia are up to 200 feet long and 10 feet in diameter at the base.

Often a novice to the intricacies of the petrification process makes the assumption that a specimen of petrified wood in a given area is of the same species as one of the trees presently growing in that vicinity. The more ancient the formation, the less likely is this to be the case. There are no *Araucaria* growing today anywhere near Petrified Forest National Park in Arizona. *Woodworthia* and *Schilderia* are extinct. Consider the location of Succor Creek, a Miocene locality that spans the Oregon/Idaho border. A 1962 study in that area identified fossilized remains of trees representing sixty-seven species, few of which occur in the area today. Most can be found growing in forests in the eastern United States in the Appalachian Mountains; others currently grow on the west slope of the Cascades; and a third set, including the *Ginkgo*, are indigenous to Asia. Likewise, while fossilized *Metasequoia* are abundant in Oregon, living specimens are located only in remote parts of China.

Petrified materials of most interest to serious paleobotanists are those that exhibit well-preserved details of the original wood structures. In some rare specimens even the internal structures of the cells are preserved. Professor Wieland used the term "over-chalcedonized" to describe petrified woods in which the cells have been disrupted or destroyed by silicification. It can certainly be observed by examining petrified materials from various formations and time periods that some have been nearly completely replaced by chalcedony while others preserve much of the original cell structure intact. One type of fossilized material that contains none of the original cell structure is a cast, which is formed when a cavity left by a decayed piece of organic material is filled by another material. This other material can be mud or sand or can be largely composed of silicas, which can produce a cast of chalcedony. In a true cast there is no evidence of cell structure or organic material.

A brief discussion of the botany of trees is appropriate here. Seed plants include gymnosperms and angiosperms. In gymnosperms seeds are borne naked in the cone; in angiosperms seeds are formed within an ovary. Gymnosperms include seed ferns, cycads, ginkgos, and conifers. Angiosperms include dicots, in which the seed germinates to two leaves and the leaves have complex net-like venation, and monocots, in which the seed germinates to one leaf (as in grasses, lilies, and palms) and the leaves have parallel veins. Gymnosperms preceded angiosperms, and dicotyledons are more primitive than monocotyledons. Conifers were the dominant large trees during the Mesozoic Era. Perhaps because of this head start, conifers still dominate as the oldest, tallest, and largest trees alive. Some bristle-cone pines live thousands of years; redwoods are our tallest and largest trees. Conifers also hold dominance in high altitudes and cold environments. Angiosperms, also known as flowering plants, are the dominant type of plant in the modern era. The origin of angiosperms is somewhat of a mystery—one which captivated the interest of Professor Wieland some time ago and which continues to interest modern researchers.

According to Professor Wieland, his study of cycads in South Dakota began with "an added wish to reach better ideas of the forests in which lived the dinosaurs." Professor Wieland became fascinated with the cycads and later with the relationship between cycads and conifers. It is clear from his works that he hoped to discover the origin of modern flowering plants, the angiosperms. Simultaneously he hoped to answer "the century old argument as to whether the cone of the pine is some sort of superflower, or was primitively an inflorescence." In *America Fossil Cycads, Volume II*, written in 1916, he wrote: "It now appears that the completer knowledge of the cycad complex is not merely requisite to a clear conception of Mesozoic plant life, but essential to an understanding of the final stages of the evolution of the higher types of seed plants." One can imagine the Professor's excitement when he learned of the nearly perfectly preserved conifers in Patagonia, Argentina. In the early 1930s he made the difficult journey to Patagonia to study this fantastic forest. He noted in his volume *The Cerro Cuadrado Petrified Forest* the incredible state of preservation of the fossilized *Araucaria* wood and cones from that forest, but had to confess: "Once more the inconclusiveness of the evidence outlining cone and flower evolution must be admitted."

The study of petrified materials may lead one to an understanding of and interest in many other subjects. Petrified fungus pockets are often noted within petrified wood. The discipline of paleoentomology is engaged when one studies fossilized wood that has been bored by insect larva, often of the flat-headed borer family. Invertebrate paleontologists study fossil woods having tunnels bored by bivalve mollusks with a long, worm-like appearance; *Teredo*, *Bankia*, and other species belong to this group which are sometimes referred to as shipworms. Several types of shipworm-bored wood from Australia and the United States contain tunnels that were filled with silica and now are composed of quartz, chalcedony, and opal.

***Note*: The several references to Dr. George Wieland and his works are not meant to overshadow the many modern advances in the study of paleobotany. Advances in science made possible by the use of the scanning electron microscope (capable of producing magnifications in excess of 100,000X), radiometric dating, and computer-enhanced modeling, as well as broader-based training in the fields of biology, geology, and statistics, have greatly enhanced this field of knowledge. However, Dr. Wieland stands as a pioneer in the field, and his place in the history of paleobotany is secure.**

2 Geology

"If the clouds be full of rain, they empty themselves upon the earth: and if the tree fall toward the south, or toward the north, in the place where the tree falleth, there it shall be."
Ecclesiastes 11:3

Paleontology and geology are tightly intertwined. Indeed, the fact that certain fossils are found consistently in specific geologic strata was critical to the development of geology. This knowledge is used to determine the relative age and identity of a formation. A subset of geology, stratigraphy, is the study of the various layers, or strata, of the earth's crust. Each stratum, or geologic formation, was deposited in a restricted geographic region under a limited set of environmental conditions. Most life-forms lived for a limited time in specialized environmental situations and could be preserved only under ideal conditions. Such conditions were rarely met, so if a stratum contains fossils, it is likely that the same stratum in another locality will yield similar fossils. For these reasons a knowledge of stratigraphy is critical to a student of fossils.

The accompanying stratigraphic chart presents the Mesozoic formations of the Henry Mountains area of southeast Utah. This area is used as an example; stratigraphic columns are available for most regions of the earth. An individual interested in exploring any region for petrified wood will benefit from a basic understanding of the formations in which petrified wood is likely to be found. This is not to suggest that petrified wood will be found in all areas where a particular formation is exposed. First, there had to have been trees. Additionally, the many conditions prerequisite for petrification had to have been present. These factors rarely coincided. Petrified wood and other petrified materials are found most consistently in certain formations and rarely, if ever, in others. Petrified wood is rarely found in the Mancos Shale Formation, the Curtis Formation, or in Wingate Sandstone. A knowledgeable prospector for petrified wood in southeast Utah would focus on the Morrison Formation, the Chinle Formation, and to some extent,

Eroded slope of Chinle Formation showing overlying Jurassic Wingate Sandstone. Grand Staircase–Escalante National Monument, Utah.

This Wingate boulder rests precariously on a column of Chinle clay holding petrified logs in Wolverine Petrified Wood Natural Area in Grand Staircase–Escalante National Monument, Utah.

the Cedar Mountain Formation. It is important to be able to recognize the appearance of the rocks that compose these formations and the manner in which they erode and form cliffs or slopes. It is also helpful to recognize the neighboring formations. A bluff of Summerville Formation sandstone may yield petrified wood eroded from the overlying Morrison Formation.

As can be seen in the accompanying photographs, the Morrison and Chinle formations are very similar in appearance. The underlying Summerville and Moenkopi formations also have an appearance similar to each other, especially in their cliff forming habits. Both the Morrison and Chinle formations contain pockets rich in petrified materials. Both consist largely of siltstone and sandstone and contain large amounts of clay derived from weathered volcanic ash. This similarity exists even though the formations were deposited some sixty million years apart and may be separated by thousands of feet of strata. These formations and similar formations in Australia, Argentina, Madagascar, Zimbabwe, and other locations in the United States and around the world are the sources for beautifully preserved petrified specimens.

Formations in any given locality may be unique to that area, may vary considerably in thickness from one place to another, and may be known by different names in different areas. Some formations exist in a lens-shape configuration, being thicker in the center of the lens and thinning out toward the edges. In addition, a formation may be subdivided into two or more members. The Chinle Formation in the southwest United States has named members that include, in descending order, Church Rock, Owl Rock, Petrified Forest, Moss Back, Monitor Butte, Shinarump, and Temple Mountain.

Araucarioxylon arizonicum **exposed in Petrified Forest National Park, Arizona.**

Typical eroded bluffs of Triassic age Chinle Formation, Grand Staircase–Escalante National Monument, Utah.

Petrified logs in Wolverine Petrified Wood Natural Area in Grand Staircase–Escalante National Monument, Utah.

Seemingly endless eroded Chinle hills in the Painted Desert area of Petrified Forest National Park.

The Crystal Forest in Petrified Forest National Park, Arizona.

Long trunk of *Araucarioxylon arizonicum* in Petrified Forest National Park, Arizona.

Petrified Forest National Park, Arizona. Millions of pounds of petrified wood from the Triassic Period of the Mesozoic Era reside here.

Petrified section of *Araucarioxylon arizonicum* in Petrified Forest National Park, Arizona.

Log sections in Petrified Forest National Park, Arizona.

ERA	PERIOD	EPOCH	END (IN YEARS BEFORE PRESENT)
CENOZOIC	Quaternary	Holocene	Today
		Pleistocene	11,000
	Tertiary	Pliocene	1,800,000
		Miocene	5,000,000
		Oligocene	25,000,000
		Eocene	36,000,000
		Paleocene	54,000,000
MESOZOIC	Cretaceous	Late	65,000,000
		Early	100,000,000
	Jurassic	Late	140,000,000
		Middle	157,000,000
		Early	178,000,000
	Triassic	Late	205,000,000
		Middle	235,000,000
		Early	241,000,000
PALEOZOIC	Permian		245,000,000
	Pennsylvanian		286,000,000
	Mississippian		325,000,000
	Devonian		360,000,000
	Silurian		410,000,000
	Ordovician		440,000,000
	Cambrian		505,000,000
PRECAMBRIAN			570,000,000
EARTH FORMED			4,500,000,000

Typical eroded bluff of Jurassic age Morrison Formation near Grand Junction, Colorado. The Wingate cliffs to the left are in Colorado National Monument.

Eroded Chinle slopes in the Painted Desert section of Petrified Forest National Park.

MESOZOIC FORMATIONS
HENRY MOUNTAINS, UTAH

PERIODS	FORMATION: DESCRIPTION	THICKNESS
Cretaceous	MANCOS SHALE: shades of gray shale, with thin lenses of sandstone and limestone.	2,000 to 3,500 feet
	DAKOTA SANDSTONE: brown to buff sandstone, some gray shale and coal.	0 - 350 feet
	CEDAR MOUNTAIN: gray to brown mudstone, sandstone, and conglomerate.	0 - 550 feet
Jurassic	MORRISON: multicolored, banded reds, purples, and green-gray clay and mudstone, conglomeratic sandstone lenses, volcanic ash.	380 - 900 feet
	SUMMERVILLE: red to red-brown siltstone and sandstone, some gypsum.	30 - 330 feet
	CURTIS: light green-gray glauconitic conglomerate, sandstone, and greenish shale.	0 - 250 feet
	ENTRADA SANDSTONE: reddish-brown siltstone and sandstone.	35 -850 feet
	CARMEL: red sandstone, green-gray to yellow-gray siltstone and shale, some limestone and gypsum.	35 - 650 feet
	NAVAJO SANDSTONE: massive, highly cross-bedded, buff to red and light gray-orange fine sandstone, lenticular limestone beds.	0 - 1100 feet
	KAYENTA: purple, red, white, tan, orange-red, and red-brown sandstone and shale.	15 - 350 feet
	WINGATE SANDSTONE: reddish-brown, massive, cliff-forming sandstone.	0 - 375 feet
Triassic	CHINLE: variegated and banded, multi-colored in grays, greens, red-browns, and purples, shales and silts, volcanic ash.	300 - 1,500 feet
	MOENKOPI: red-brown siltstone and sandstone, dominantly red beds.	0 - 800 feet

Weathered petrified log in Petrified Forest State Park, Escalante, Utah.

Gnarly, lichen-covered log in Petrified Forest State Park, Escalante, Utah.

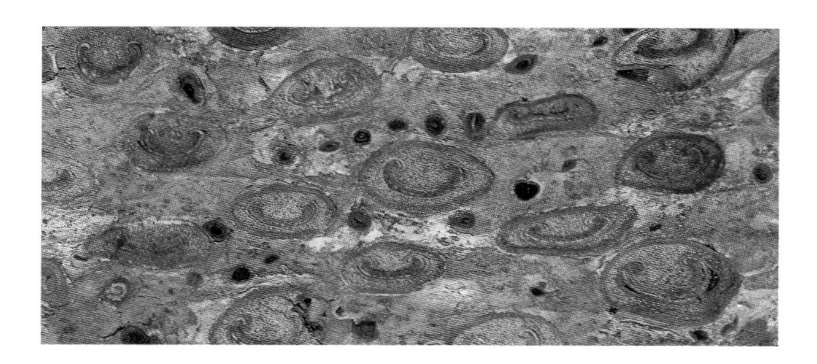

3 Mineralogy

"Beauty is truth, truth is beauty – that is all Ye know on earth, and all ye need to know.

From *Ode on a Grecian Urn* John Keats 1795-1821

Petrified wood was a plant and is now largely a mineral. Because we are dealing with minerals, some understanding of mineralogy is helpful.

The basic substances composing the bulk of petrified woods are hydrous microcrystalline varieties of quartz that can be generally termed as chalcedony or opal. Chalcedonic material can be agate and its varieties onyx and sardonyx, when banded or variegated; carnelian when it attains a bright orange to red translucency; prase or plasma in greenish colors; and jasper when subtranslucent to opaque. Opal almost always occurs as "common opal." "Precious opal," which shows the beautiful play of colors, rarely replaces fossil wood. Crystalline quartz varieties such as amethyst, smoky quartz, and citrine are found in limb casts or lining cavities of fossilized wood.

Chalcedony consists of microcrystalline aggregates or bundles of silica fibers that can adsorb up to ten percent water. Hydroxyl ions may be present between the interlocking fibers. This interlocking fibrous nature gives chalcedony its toughness, similar to the structure of jade. In agates, the microscopic fibers grow orthogonal to the banding and can exhibit very regular twisting. If the banding is sufficiently fine (400 to 17,000 layers per inch) and the twisting of fibers uniform, a diffraction grating that breaks light rays into its component colors may be created. This results in the well known iris or rainbow agates, a phenomenon best demonstrated when the agate is cut into a thin slice 1/16 to 1/8 inch thick and held up to a light. Limb casts from Wiggins Fork, Wyoming, and other locations demonstrate the iris effect.

Opal is a colloidal silica that has a waxier luster than chalcedony, usually contains more water, and is not as hard or tough. Common opal is translucent milky white to yellowish, greenish, and grayish and often permineralizes limbs with faithful preservation

Wyoming [Wiggins Fork]
Cenozoic; Tertiary
unidentified [iris effect]
Dayvault 15.5 cm

The woody cell walls in this Jurassic age branch from Yellow Cat, Utah, have been completely replaced by carnelian and blue chalcedony.
Utah, USA [Yellow Cat] Mesozoic; Jurassic
unidentified fm: Morrison
Daniels 5 cm

One end of this branch section formed a cluster of quartz crystals.
Wyoming, USA [Blue Forest]
Cenozoic; Tertiary; Eocene
unidentified fm: Green River
Jones 12 cm

Polished cross-section of the specimen in the previous photograph (*BOTTOM LEFT*), revealing a combination of permineralization and crystallization.
Wyoming, USA [Blue Forest]
Cenozoic; Tertiary; Eocene
unidentified fm: Green River
Jones 7.5 cm

of the original wood texture. Precious opal forms as replacements in limb casts and is occasionally associated with common opal permineralization of the same limb. The difference between common and precious opal is the packing of tiny (150 to 300 nanometers in diameter) amorphous silica spheres that constitute all opal. If the spheres are evenly sized and regularly packed, as in precious opal, they can act as a diffraction grating and break white light into its component colors. If, however, the spheres are of unequal size and not evenly packed, as in common opal, no diffraction grating is developed. For precious opal formation, groundwater conditions must be extremely stable to allow the colloidal spheres of silica to settle out of solution and arrange themselves into an orderly, crystal-like array.

When describing the process of silicification, the term agatize is often used by collectors when it might be more accurate to use the term silicify or chalcedonize. Many materials referred to as agatized are actually jasperized or opalized. In many instances the process involves a combination of these mineralizations. Petrified wood from Nevada, for example, is often a combination of opalized, jasperized, and chalcedonized.

A mineral's hardness depends on its atomic structure. Hardness is defined as the resistance a smooth surface offers to abrasion or scratching. A widely used scale for measuring hardness is the Mohs scale, named for Austrian mineralogist Frederick Mohs who first published this scale in 1822. The Mohs scale ranks ten known minerals in ascending order of relative hardness:

1. Talc
2. Gypsum
3. Calcite
4. Fluorite
5. Apatite
6. Orthoclase
7. Quartz
8. Topaz
9. Corundum
10. Diamond

The actual degree of increase in hardness does not progress uniformly from one number to the next. Diamonds are vastly harder than corundum, while fluorite is only slightly harder than calcite. Another measure for the nature of minerals is toughness. Opal is a brittle mineral, while chalcedony has a tough, fibrous quality. Opal has a hardness ranging from 5½ to 6½ on the Mohs scale; chalcedony and jasper are at the upper end of this range.

The many colors found in petrified materials are a function of groundwater chemistry during formation. How a tree becomes petrified, the location, and the composition of the overlying strata all influence the eventual color of the petrified material. Consider, for example, the trees in Arizona's Chinle Formation which became buried in a vast flood plain under layers of fluvial sediments more than 2,500 feet thick in Upper Triassic sediments alone. These sediments consisted largely of ash-laden silts, clays, and sands. Acidic solutions containing iron and manganese in various states of oxidation and reduction created petrified woods in a rainbow of colors. Petrified woods from other locations have their own distinctive colorations. Those from the Blue Forest in Wyoming are generally shades of brown with blue chalcedony. Those from Zimbabwe are generally shades of brown with much green, apparently from chromium, although green color in petrified wood is usually created by ferrous oxides. Woods from Washington, having been directly buried by lava flows, are often subtle yet attractive shades of gray and brown. Some of the colors found in petrified wood and corresponding minerals are found in Table 1.

Table 1
Colors found in petrified wood and some corresponding minerals

Color	Minerals
Red	iron (ferric)
Orange	iron (ferric)
Yellow	iron (ferric), uranium
Green	iron (ferrous), copper, cobalt, chromium, uranium, nickel
Blue	copper, manganese, cobalt, chromium
Violet	manganese, iron (ferric)
Purple	iron (ferric), manganese
Brown	iron (ferric), uranium
Black	manganese, carbon, iron (ferric)
White	silicon dioxide
Gray	silicon dioxide

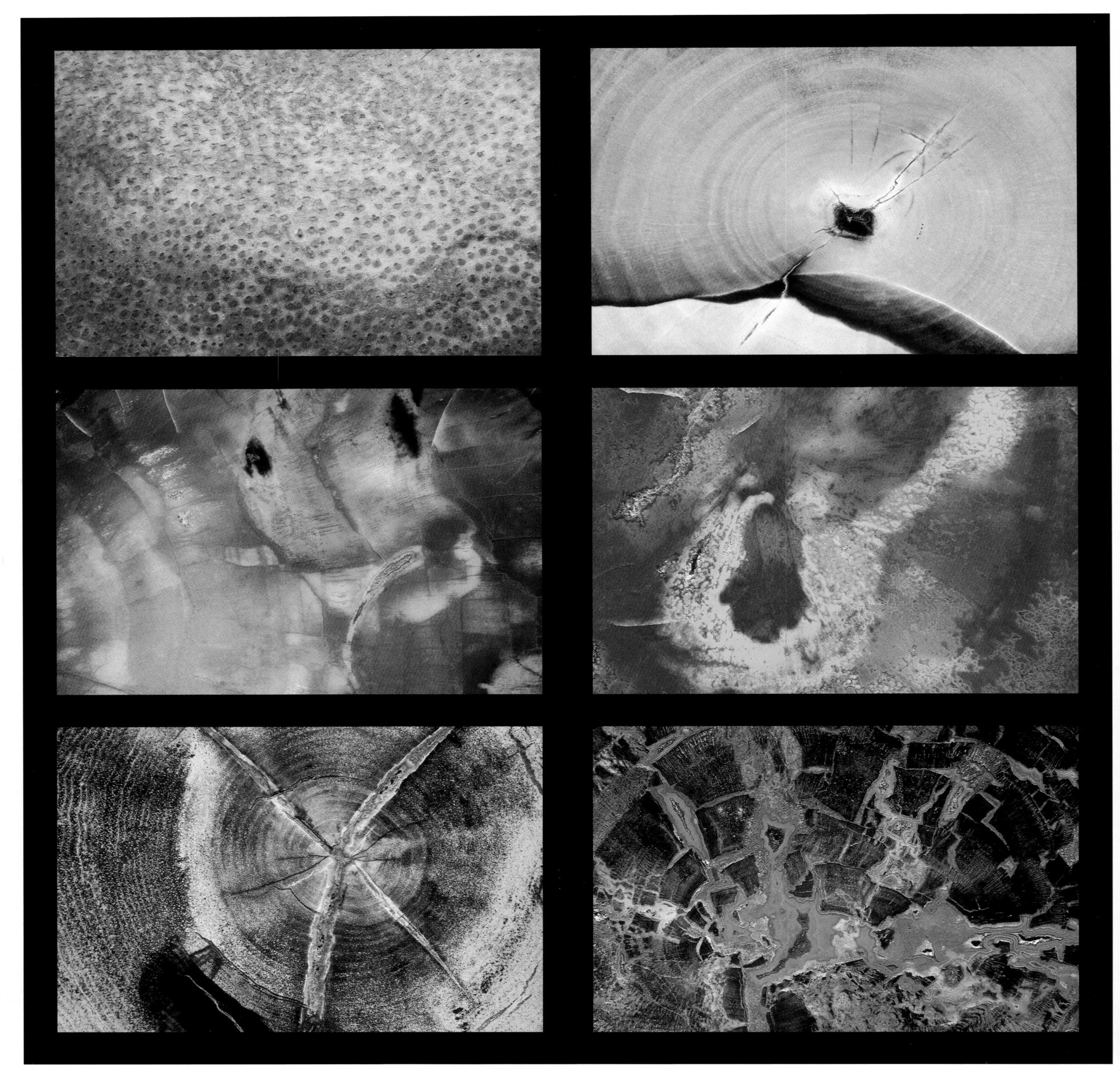

Louisiana, USA
Cenozoic; Tertiary; Oligocene
Palmoxylon **fm: Catahoula**
Akin

Washington, USA
Cenozoic; Tertiary; Miocene
Sequoia **[redwood]**
Daniels

Arizona, USA
Mesozoic; Triassic
Araucarioxylon arizonicum **fm: Chinle**
Daniels

Arizona, USA
Mesozoic; Triassic
Araucarioxylon arizonicum **fm: Chinle**
Daniels

Oregon, USA [Deschutes]
Cenozoic; Tertiary; Miocene/Pliocene
Carya **[hickory]**
fm: Columbia River Basalt
Daniels

Crystal pockets of quartz and amethyst within carnelian fortifications in a heavily fungus-rotted *Araucaria* log.
Patagonia, Argentina **Mesozoic; Jurassic**
Araucaria
Daniels

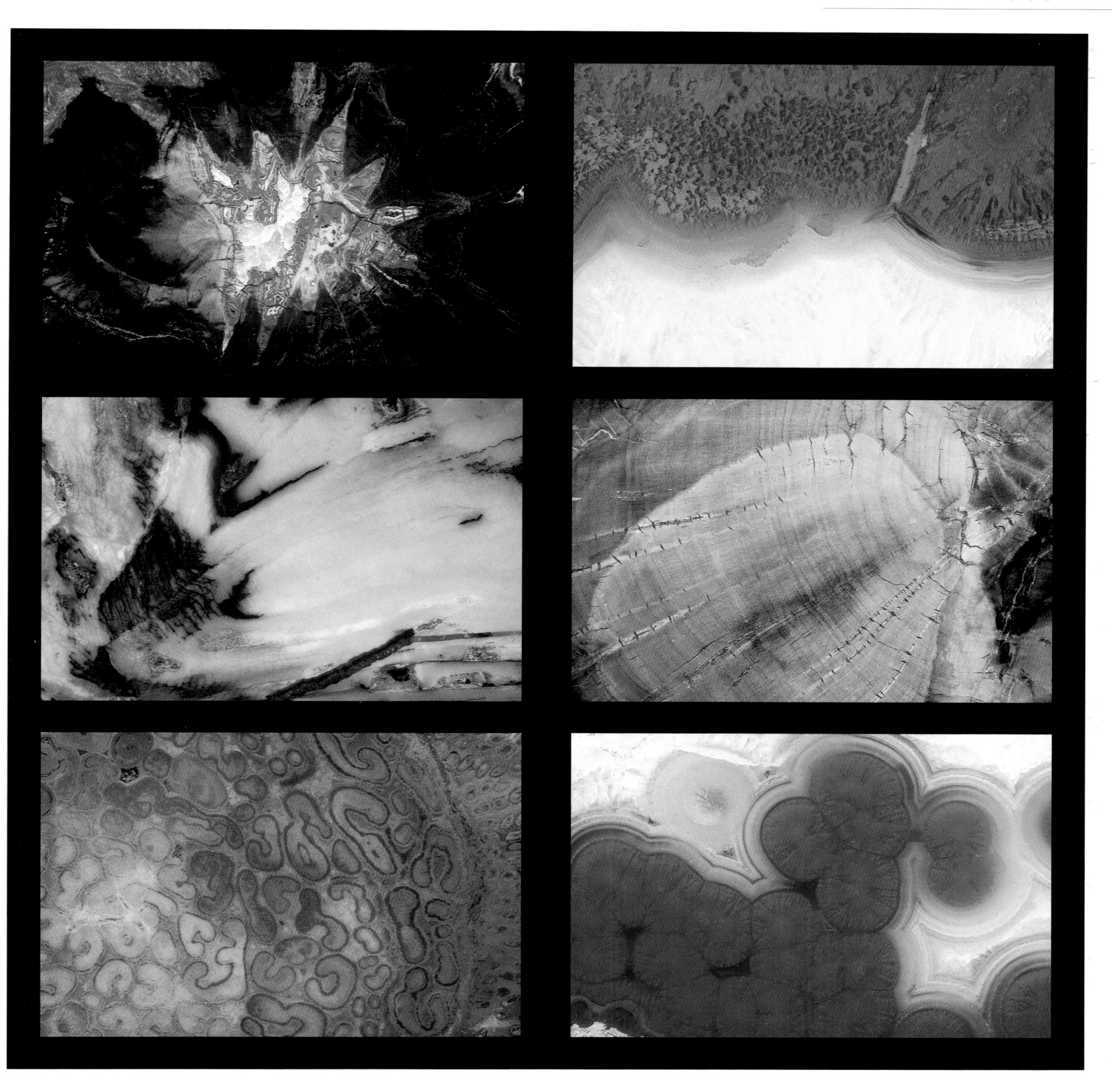

Arizona, USA
Mesozoic; Triassic
***Dadoxylon* star   fm: Chinle**
Daniels

Utah, USA [Yellow Cat "redwood"]
Mesozoic; Jurassic
unidentified   fm: Morrison
Daniels

Nevada, USA [Hubbard Basin]
Cenozoic; Tertiary
unidentified
Daniels

Arizona, USA
Mesozoic; Triassic
***Woodworthia*   fm: Chinle**
Daniels

Maranhao, Brazil [Araguaina]
Paleozoic; Permian
Tietea singularis
Daniels

Utah, USA [Yellow Cat "redwood"]
Mesozoic; Jurassic
unidentified   fm: Morrison
Kladder

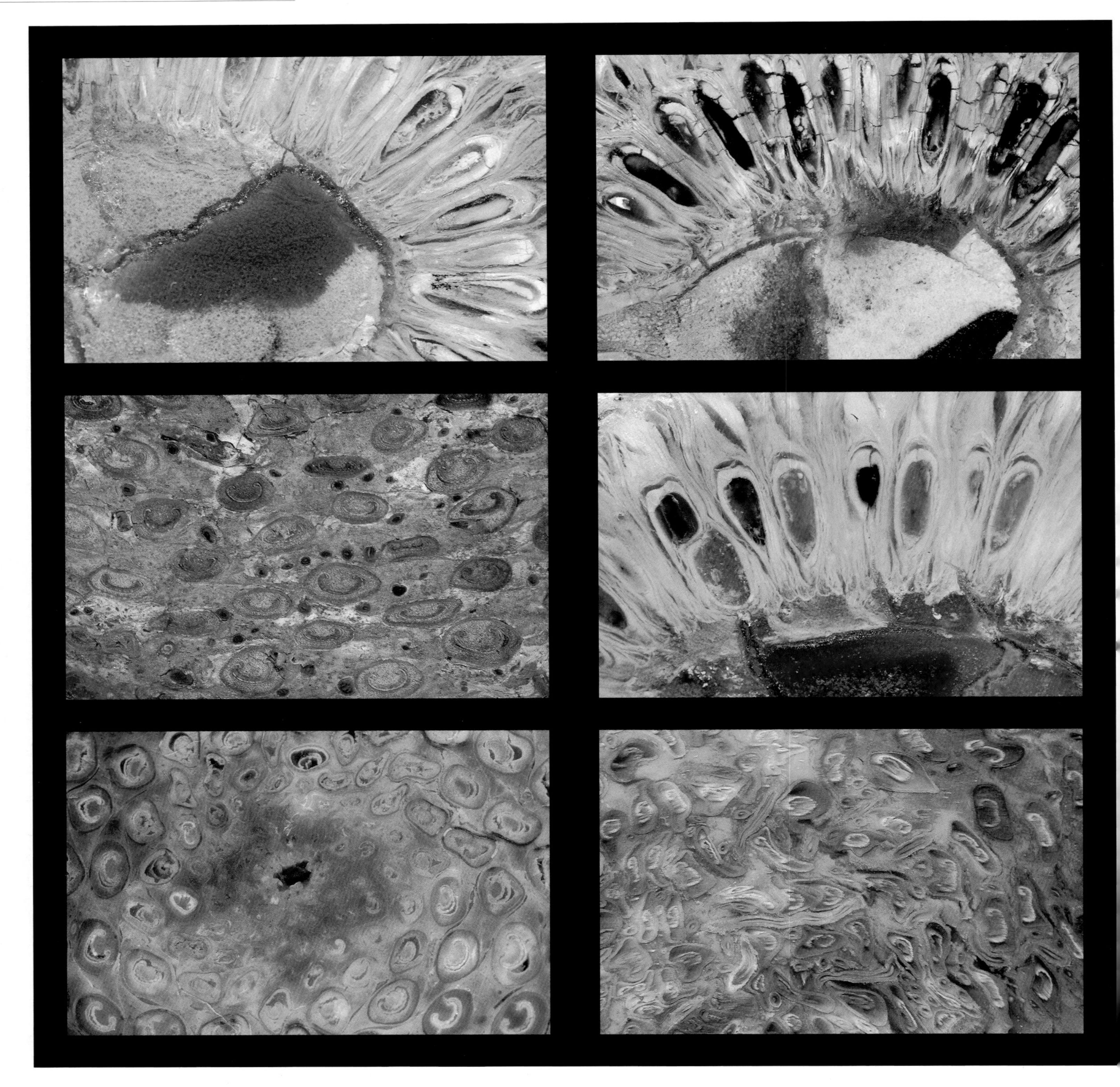

Patagonia, Argentina
Mesozoic; Jurassic
***Araucaria mirabilis* cone**
Daniels

Queensland, Australia
Mesozoic; Jurassic
***Osmundacaulis* fm: Miles**
Daniels

Queensland, Australia
Mesozoic; Jurassic
unidentified fern fm: Miles
Rigel

Patagonia, Argentina
Mesozoic; Jurassic
***Araucaria mirabilis* cone**
Daniels

Patagonia, Argentina
Mesozoic; Jurassic
***Araucaria mirabilis* cone**
Daniels

Queensland, Australia
Mesozoic; Jurassic
unidentified fern fm: Miles
Rigel

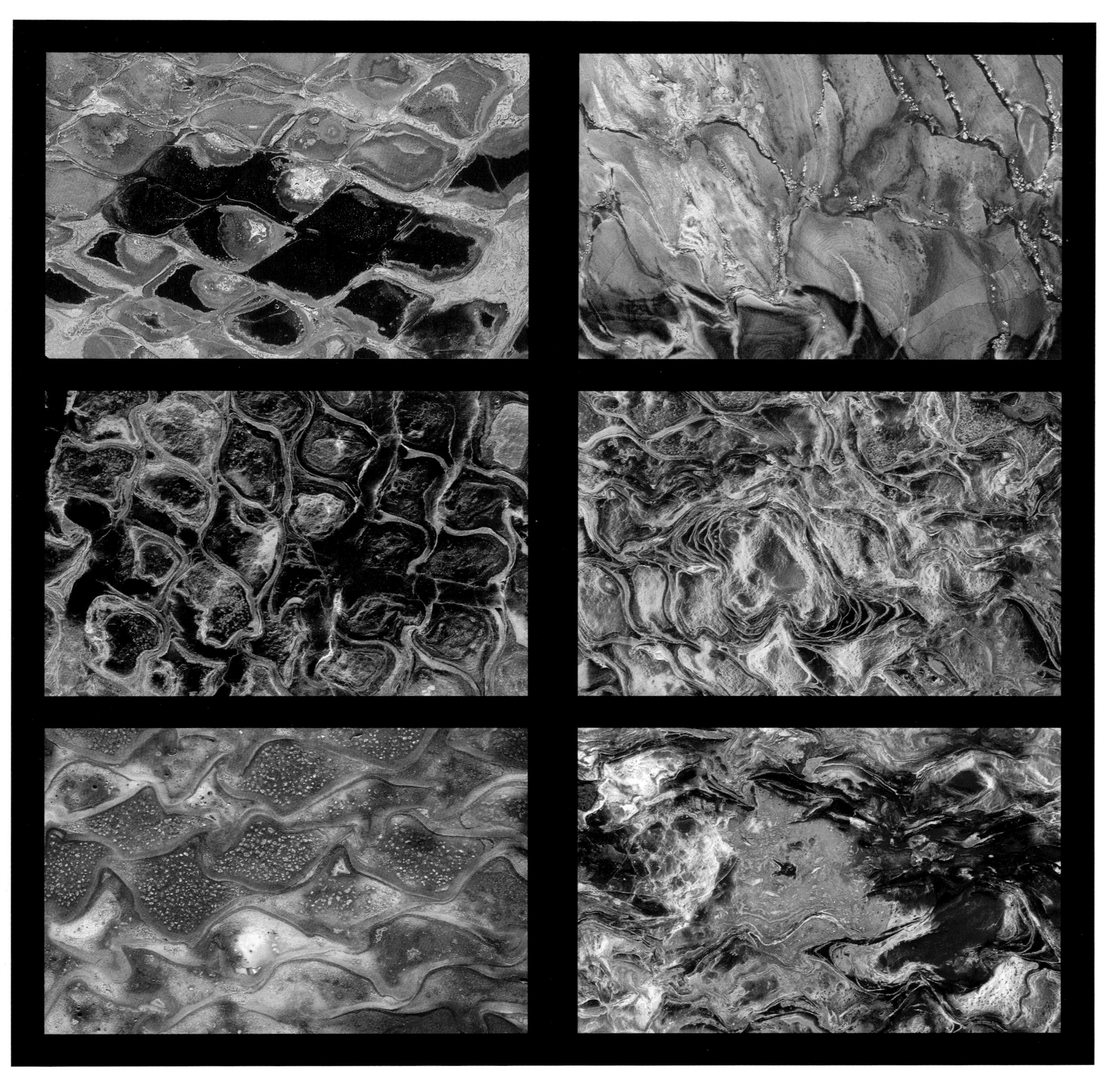

Utah, USA [Henry Mountains]
Mesozoic; Jurassic
Cycadeoidea **fm: Morrison**
Branson

Queensland, Australia
Mesozoic; Jurassic
"Pentoxylon" **fm: Miles**
Daniels

Patagonia, Argentina
Mesozoic; Jurassic/Cretaceous
Cycadeoidea
Daniels

Patagonia, Argentina
Mesozoic; Jurassic/Cretaceous
Cycadeoidea **[cone in center]**
Daniels

Patagonia, Argentina
Mesozoic; Jurassic/Cretaceous
Cycadeoidea
Daniels

Patagonia, Argentina
Mesozoic; Jurassic/Cretaceous
Cycadeoidea **[cone in center]**
Daniels

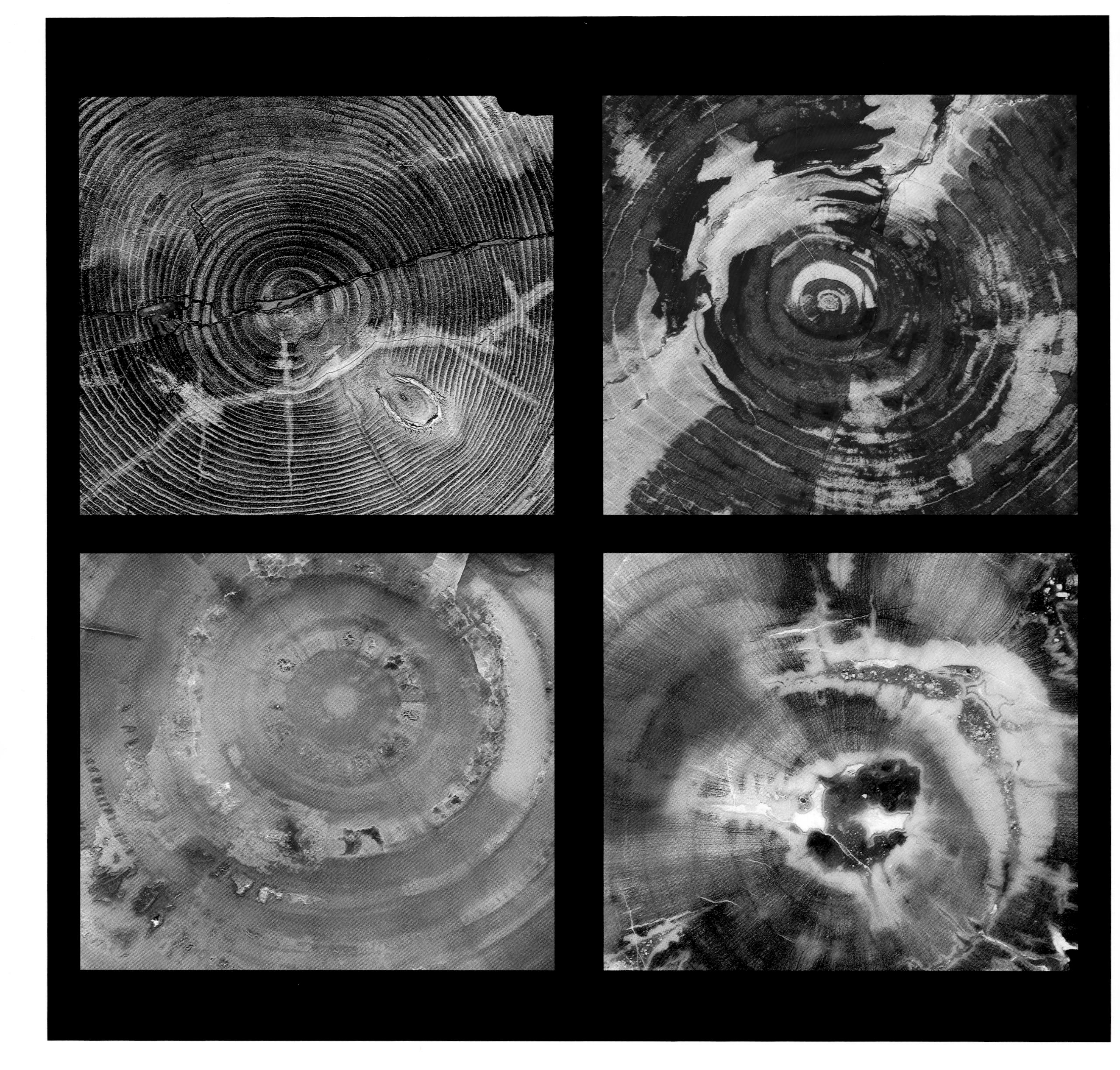

Oregon, USA [Sweet Home]
Cenozoic; Tertiary; Oligocene/Miocene
unidentified
Daniels

Queensland, Australia
Mesozoic; Jurassic
conifer fm: Miles
Daniels

Arizona, USA
Mesozoic; Triassic
conifer fm: Chinle
Daniels

Utah, USA [Henry Mountains]
Mesozoic; Jurassic
conifer fm: Chinle
Daniels

Collecting Petrified Materials

4

"No motion has she now, no force; She neither hears nor sees; Rolled round in earth's diurnal course, With rocks, and stones, and trees."

From *A Slumber did my Spirit Seal* William Wordsworth 1770 - 1850

It is first important to note that collection of petrified materials from virtually all land is subject to law. Petrified materials on private land are protected by laws dealing with trespass and theft. Landowner permission must be obtained before entering private property to collect. All state and federal lands in the United States also are subject to collecting laws. Other countries have their own laws dealing with the collection of petrified materials. Before collecting, check with the local office of the responsible agency. Collection of rocks or fossils is forbidden in national parks and monuments in the United States and on Native American lands. One may collect petrified wood on much of the land administered by the Bureau of Land Management (BLM), subject to local rules and regulations. A pamphlet distributed by the BLM in Western Colorado contains the following statement:

> *Petrified wood: You may pick up 25 pounds and one piece of petrified wood per day, not to exceed 250 pounds per year. For example, on June 1, 1994, you pick up fifteen small pieces that weigh a total of 25 pounds, plus one large petrified branch that weighs 100 pounds. On June 2, 1994, you pick up ten more pieces that weigh a total of 25 pounds and one more petrified branch that weighs 100 pounds. You have now picked up your limit of 250 pounds and you cannot collect any more petrified wood until June 1, 1995. It is not legal for you to sell the petrified wood you have collected on public lands.*

This, of course, is subject to change without notice.

Some statutes pertaining to the collection of petrified wood on federal lands are included in the

Utah, USA [Henry Mountains]
Mesozoic; Jurassic
conifer fm: Morrison
Jones 9 cm tall

Madagascar
Mesozoic; Triassic
conifer
Daniels 3.5 cm tall

Patagonia, Argentina
Mesozoic; Jurassic
Araucaria
Daniels 9 cm tall

Utah, USA [Henry Mountains]
Mesozoic; Jurassic
conifer fm: Morrison
Daniels 9.5 cm tall

Spheres cut from petrified wood from Argentina, Madagascar, Arizona, Louisiana, and Oregon.

Appendix of this volume. The definition of petrified wood as found in Title 30 of the United States Code is:

> *Petrified wood as used in this Act means agatized, opalized, petrified, or silicified wood, or any material formed by the replacement of wood by silica or other matter. 30 U.S.C. Sec. 611.*

It is appropriate that any unusual find of petrified materials be brought to the attention of a professional paleobotanist at a local museum, university, or governmental agency. Many of the finest specimens now in museums and research collections of many universities were at one point in the home of a private collector. Collecting by individuals is important to the process whereby the best and most important specimens are funneled to public institutions. Therefore, it is important for private collectors to keep a record of each specimen and to include as much locality and geologic information as possible.

Individuals interested in petrified wood, cones, ferns, and cycads should visit preserves such as Petrified Forest National Park and famous collecting localities. A sure manner of collection is to purchase specimens from a reputable fossil dealer. Every year there are hundreds of gem and mineral shows. The largest shows are in Denver, Colorado, in September, Munich, Germany, in November, and Tucson, Arizona, in February. Additionally, there are thousands of rock shops and other businesses that sell good specimens. It would take an individual a lifetime as well as an unlimited travel budget to personally collect the petrifications available at a major show or within the inventory of large dealers.

Specimens sought for a collection are a matter of personal preference. To many collectors, a prized specimen of petrified wood is one that is full round, that is, a specimen that encompasses the entire perimeter of a branch or a trunk and contains a full set of annual rings. A top quality specimen has no significant fractures, has attractive color, and is completely silicified (chalcedonized, jasperized, and/or opalized). There is a preference for chalcedony that displays deep translucence. Specimens with enough of the natural exterior to give a flavor of the outer surface are desirable, as are those with attractive knots or small branches. Some collectors seek specimens that are double-hearted or multiple-hearted, meaning that the polished cross-section contains more than one set of annual rings, having been a fork in a trunk or branch. While some people collect by specimen, others collect by type, for example, by attempting to obtain a representative specimen of every genus of tree known to have fossilized in a locality. In general, smaller specimens are collected as small trunks or limb sections and larger specimens as cross-sections or "slabs." Some collectors eschew specimens which are so "over-chalcedonized" so as to elude identification of the wood; others relish attractive casts. These same factors pertain to the collection of petrified ferns. Cones are collected as whole specimens or are cut and polished. Cycads are collected as whole specimens (although these are extraordinarily rare) or in sections containing the outer armor. Tangential sections of the outer armor polished to reveal the diamond patterns are prized by many collectors.

Cutting or polishing silicified wood requires a substance harder than quartz. Petrified wood is generally cut with a diamond saw blade cooled with oil or water. Some larger petrified logs are cut with an abrasive wire in a slurry of corundum powder and water. Corundum and diamond are also generally used to polish silicified materials. For the final polish, other very fine, very hard powders, such as tin oxide and cerium oxide, are used. A properly cut and polished specimen will not be coated in any way; the actual quartz (chalcedony, jasper, and/or opal) surface, brought to a lustrous polish, is what you see. The specimens pictured in this volume have had nothing added to the polished surfaces.

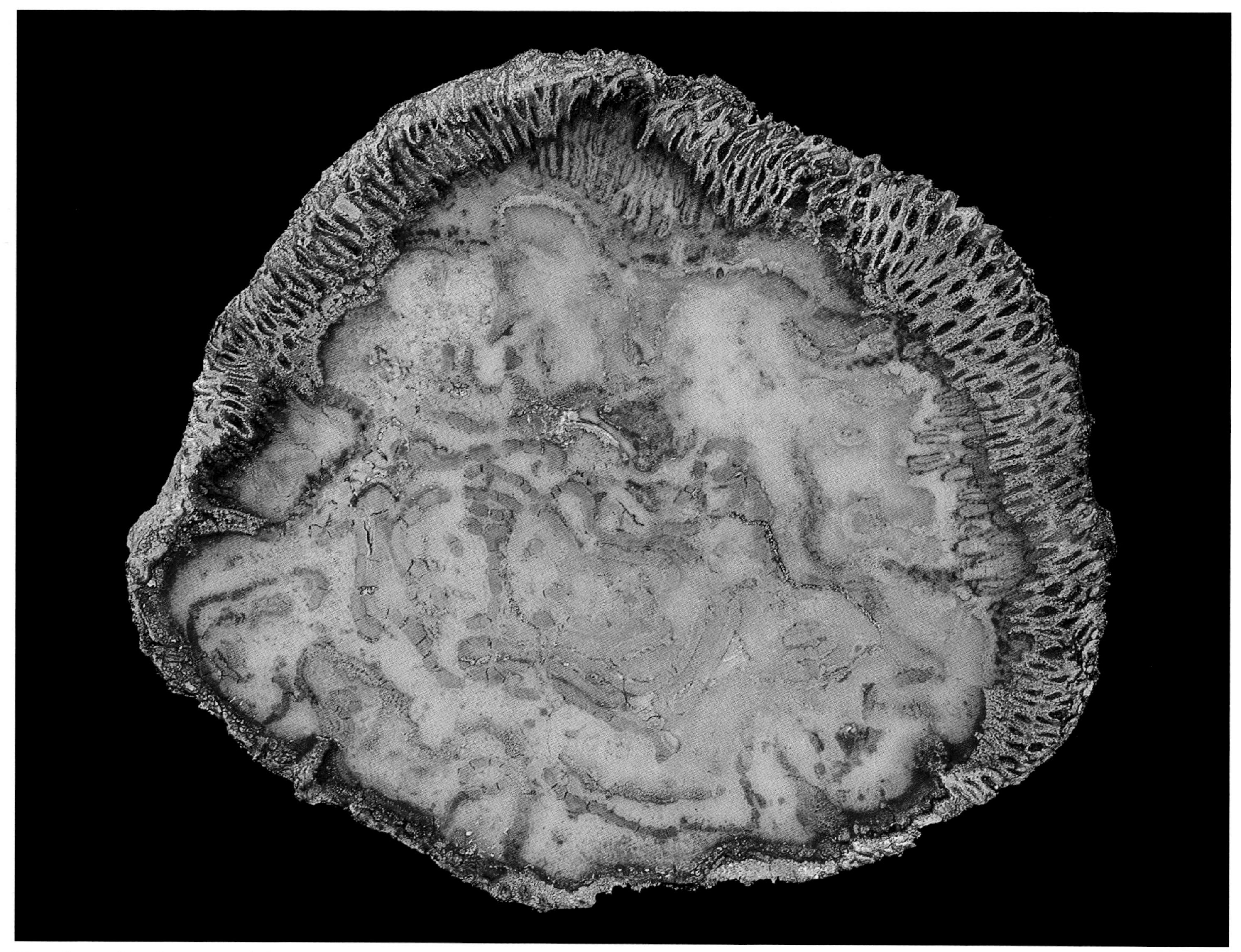

Tree fern of unknown origin.
Daniels 14 cm

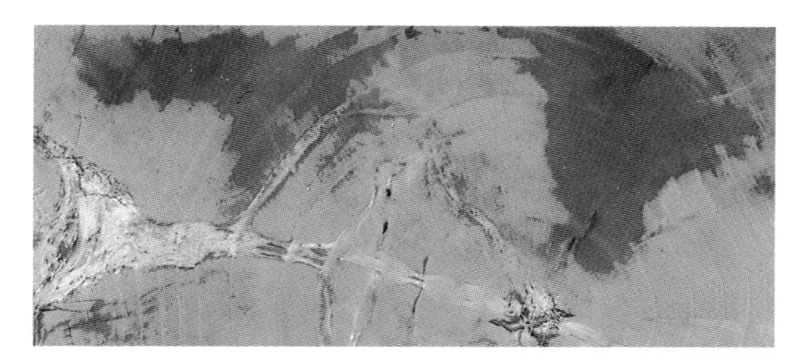

Petrified Wood From Around the World 5

ARGENTINA

In the 1920s the first published studies on the petrified wood and cones of Patagonia appeared in Europe and the United States. Researchers traveled to the area to collect specimens and conduct further studies. Materials collected by paleontologists from the Field Museum in Chicago inspired Professor Wieland of Yale University to make the journey. Petrified wood from Patagonia is often well silicified and is known for excellent exterior detail and a general lack of fractures. Petrified *Araucaria* branches often are found in close proximity to the cones, occasionally with cones still attached to branches.

The majority of the petrified specimens from Argentina are from the Jurassic equivalent to the Morrison Formation in the United States. Many Argentine petrified materials are preserved in national parks, such as the Bosque Petrificado Cerro Cuadrado in Patagonia.

"That this one forest should remain so unparalleled is due to nearly pure chance and type."
Professor G.R. Wieland, 1935

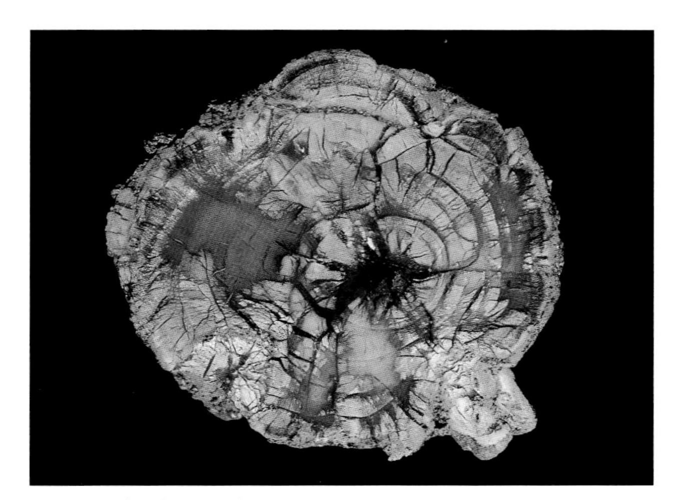

Patagonia, Argentina
Mesozoic; Jurassic
Araucaria
Daniels 6.5 cm

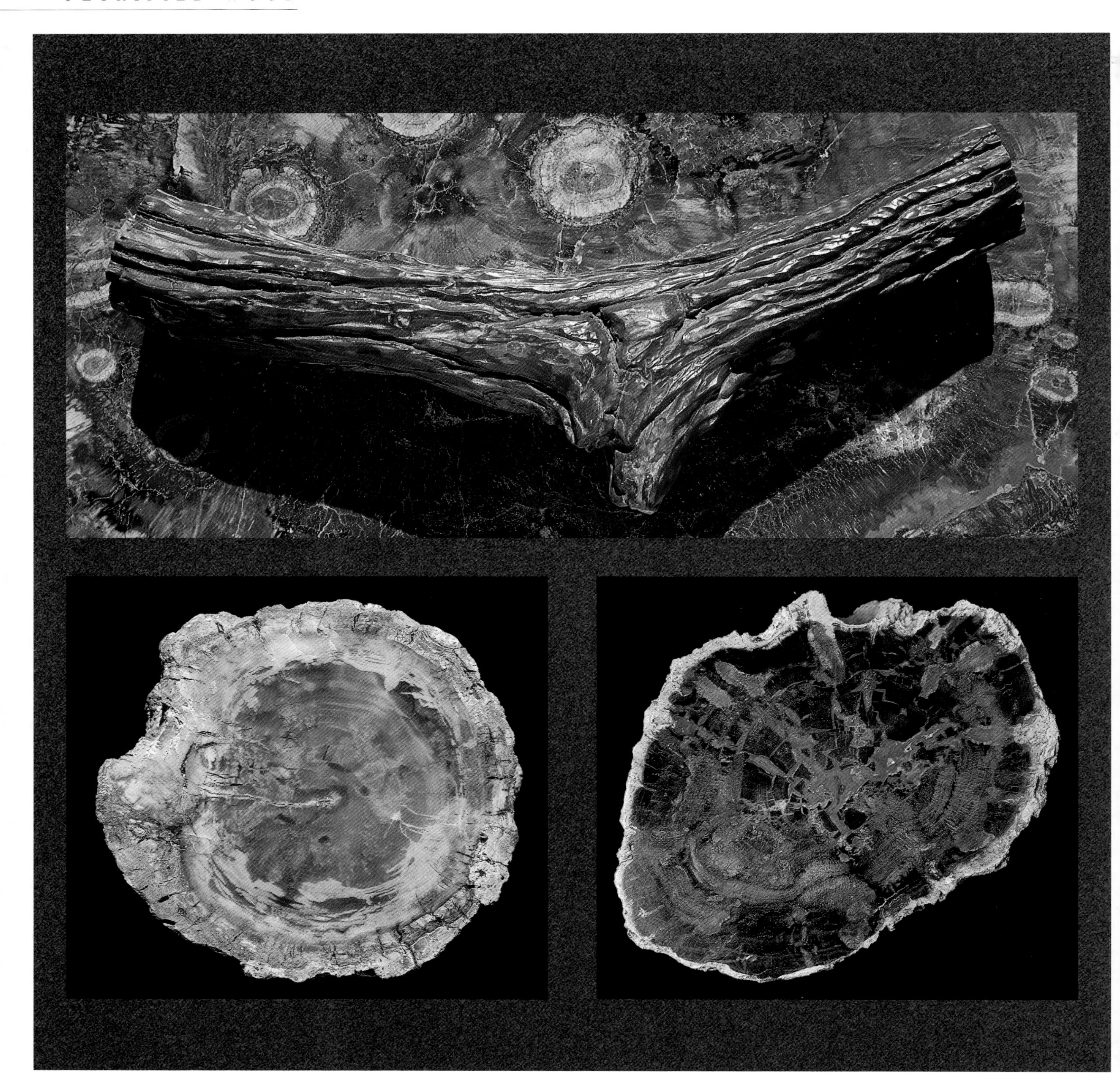

Gnarly, wind-polished Triassic conifer branch. Background is a cross-section of a fungus-pocketed Triassic conifer from Arizona.
Patagonia, Argentina
Mesozoic; Triassic
conifer
Daniels 44 cm

Patagonia, Argentina
Mesozoic; Jurassic
Araucaria
Daniels 10 cm

Patagonia, Argentina
Mesozoic; Jurassic
conifer
Daniels 16.5 cm

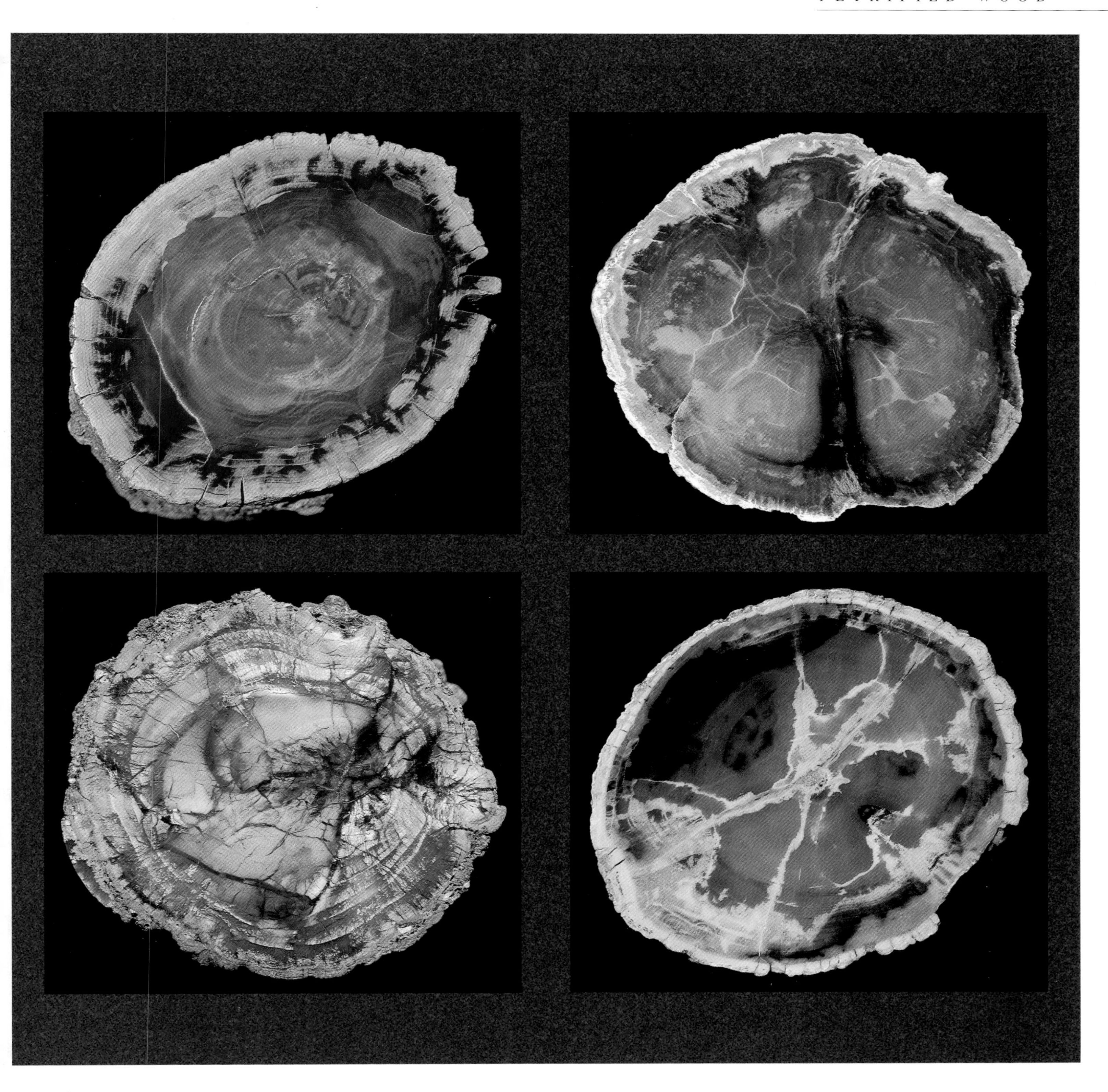

Patagonia, Argentina
Mesozoic; Jurassic
Araucaria
Branson 12.5 cm

Patagonia, Argentina
Mesozoic; Jurassic
Araucaria
Daniels 6 cm

Patagonia, Argentina
Mesozoic; Jurassic
Araucaria
Daniels 6 cm

Patagonia, Argentina
Mesozoic; Jurassic
Araucaria
Daniels 8 cm

Patagonia, Argentina
Mesozoic; Jurassic
Araucaria
Daniels 5.5 cm tall 9.5 cm

Patagonia, Argentina
Mesozoic; Jurassic
Araucaria
Branson 9 cm tall 11 cm

Patagonia, Argentina
Mesozoic; Triassic
conifer
De Los Santos 10 cm

Patagonia, Argentina
Mesozoic; Jurassic
Araucaria
Daniels 6.5 cm tall 7.5 cm

Patagonia, Argentina
Mesozoic; Jurassic
Araucaria
Daniels 4.5 cm tall 8 cm

Patagonia, Argentina
Mesozoic; Jurassic
Araucaria
Daniels 5.5 cm

Patagonia, Argentina
Mesozoic; Jurassic
Araucaria
Daniels 23.5 cm

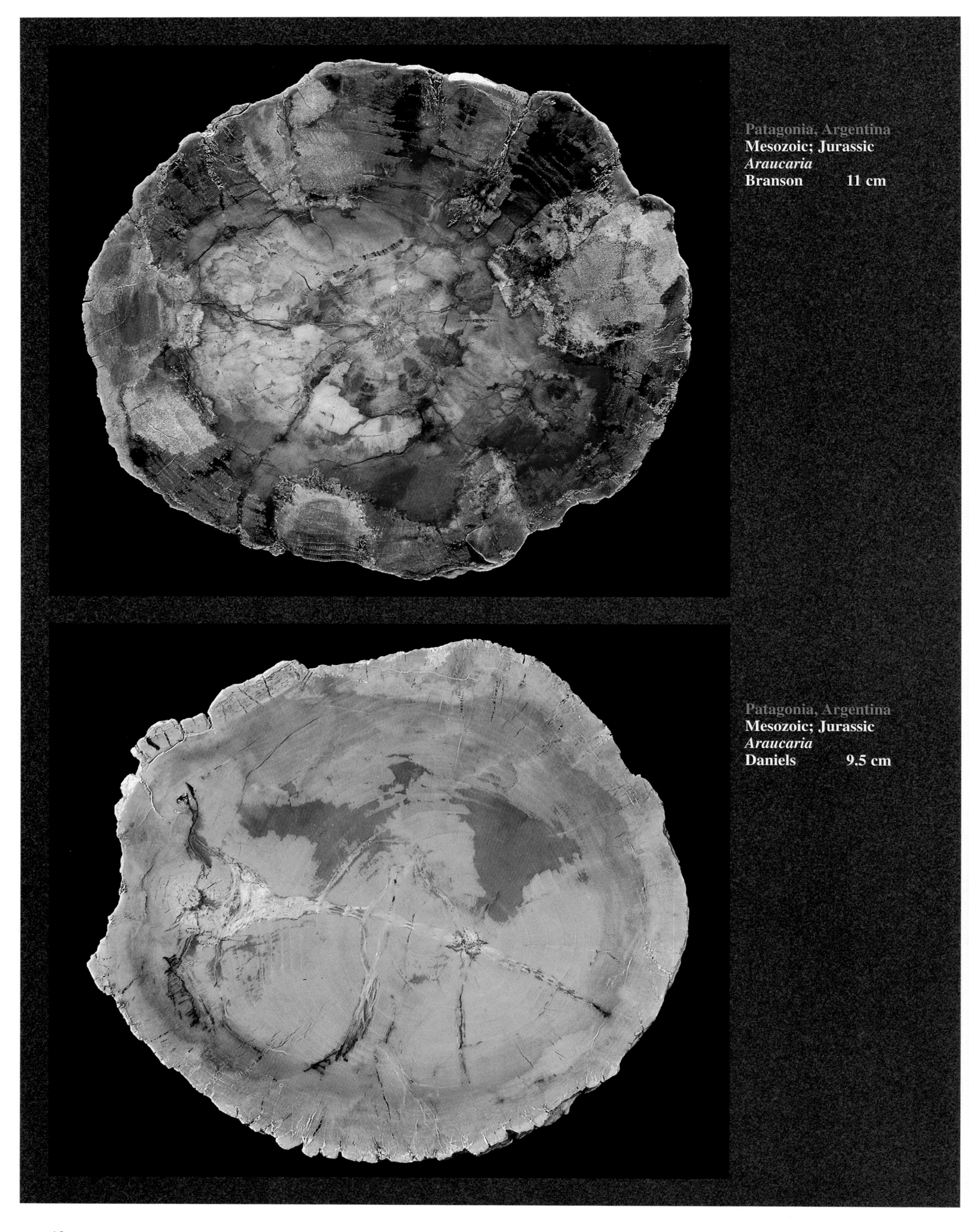

Patagonia, Argentina
Mesozoic; Jurassic
Araucaria
Branson 11 cm

Patagonia, Argentina
Mesozoic; Jurassic
Araucaria
Daniels 9.5 cm

Patagonia, Argentina
Mesozoic; Jurassic
Araucaria
Daniels 18 cm

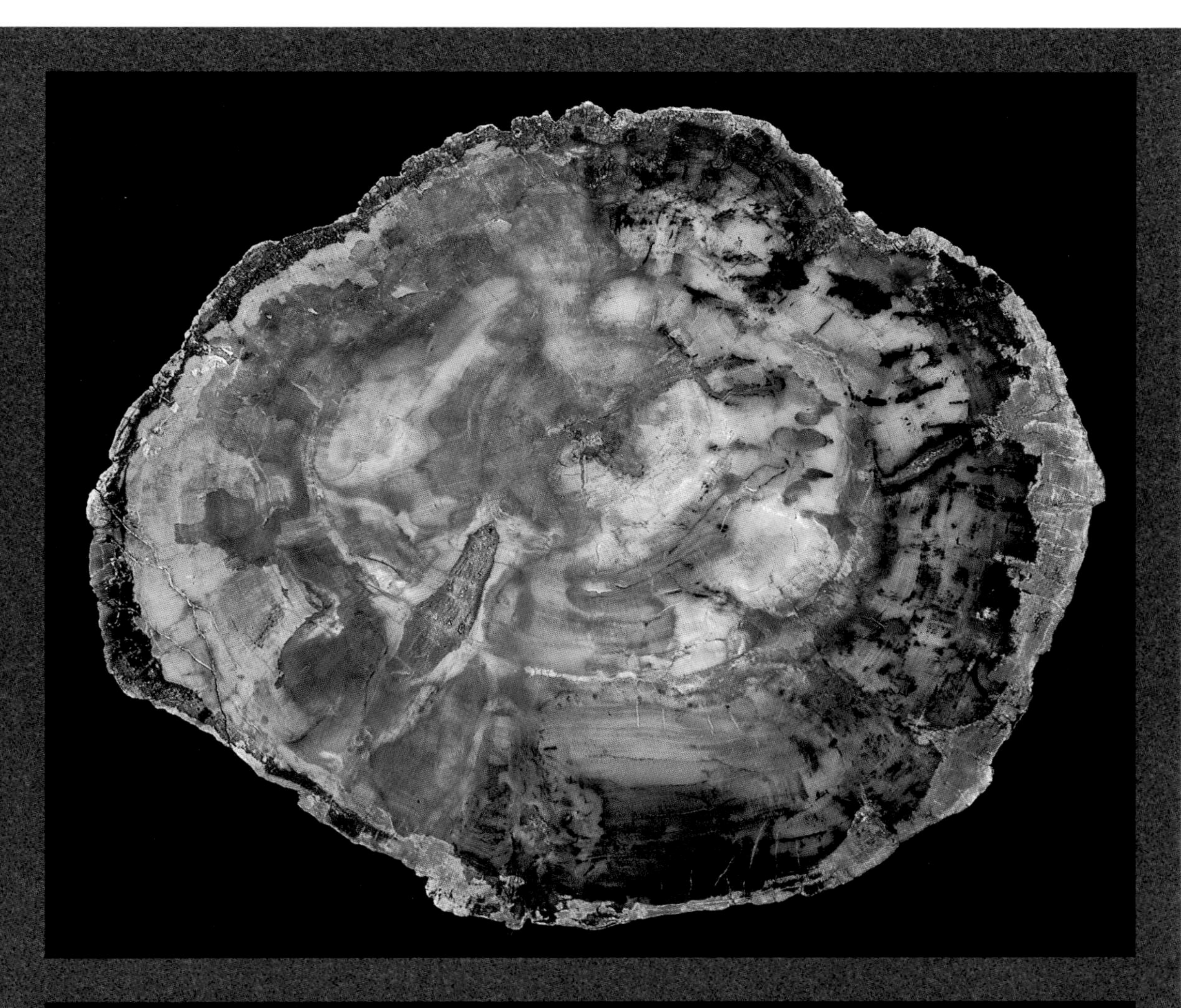

Patagonia, Argentina
Mesozoic; Cretaceous
unidentified
Daniels 23.5 cm

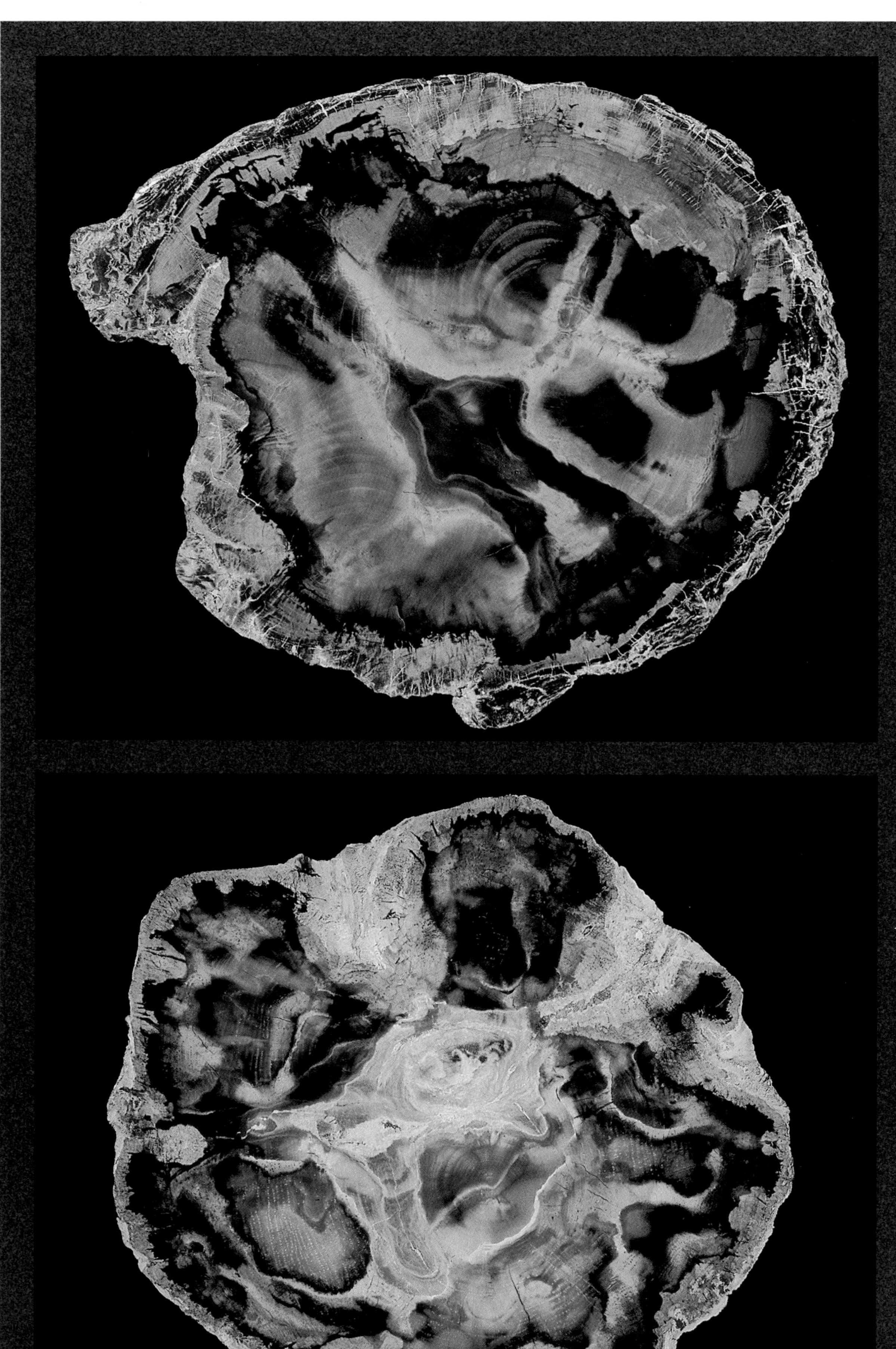

Patagonia, Argentina
Mesozoic; Jurassic
***Araucaria* [double-hearted]**
Branson 15.5 cm

Patagonia, Argentina
Mesozoic; Jurassic
Araucaria
Nottingham 30 cm

AUSTRALIA

Petrified wood from Queensland, Australia, usually bears evidence of river transport and wear. Surface detail is indistinct, but cell structure is excellent. These specimens are solidly jasperized in a wide range of colors. Shipworm wood is generally opalized. The most commonly obtained specimens are from the Jurassic age Miles Formation.

Queensland, Australia
Mesozoic; Jurassic conifer
fm: Miles
Daniels
22.5 cm

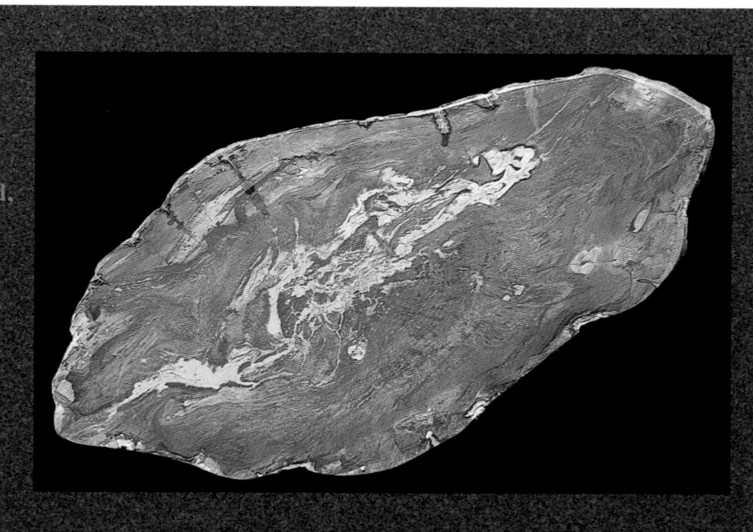

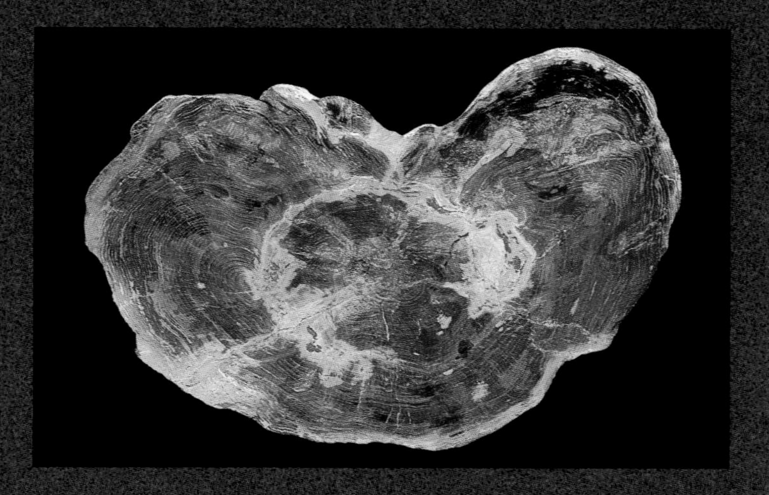

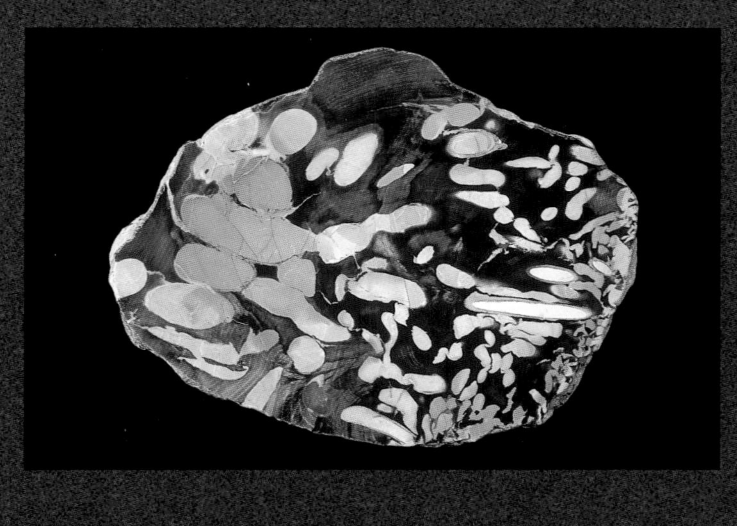

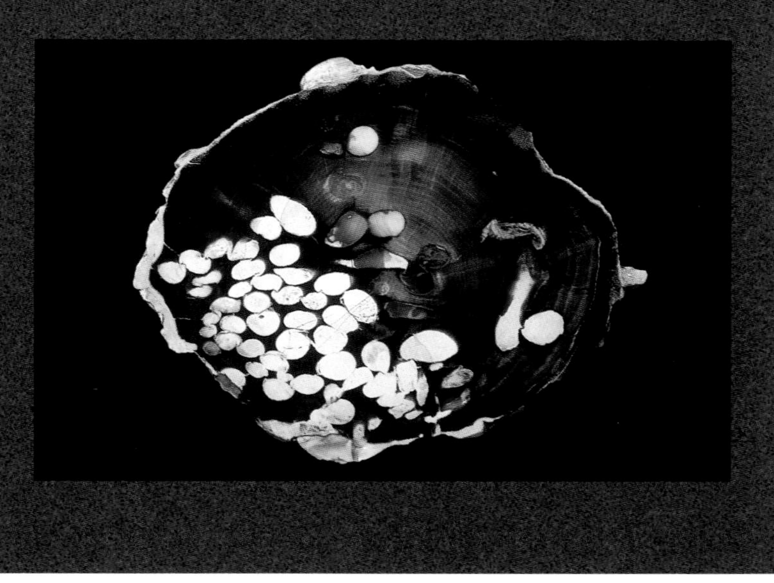

Queensland, Australia
Mesozoic; Jurassic
conifer
Daniels 8 cm

Queensland, Australia
Mesozoic; Jurassic
conifer [fire-scarred] fm: Miles
Daniels 29.5 cm

Australia
Mesozoic; Jurassic
conifer with mollusk bore tunnels
Daniels 18 cm

Australia
Mesozoic; Jurassic
conifer with mollusk bore tunnels
Daniels 12 cm

Queensland, Australia
Mesozoic; Jurassic
conifer fm: Miles
Daniels 17.5 cm

Queensland, Australia
Mesozoic; Jurassic
conifer fm: Miles
Daniels 24 cm

Queensland, Australia
Mesozoic; Jurassic
conifer fm: Miles
Lark 16.5 cm

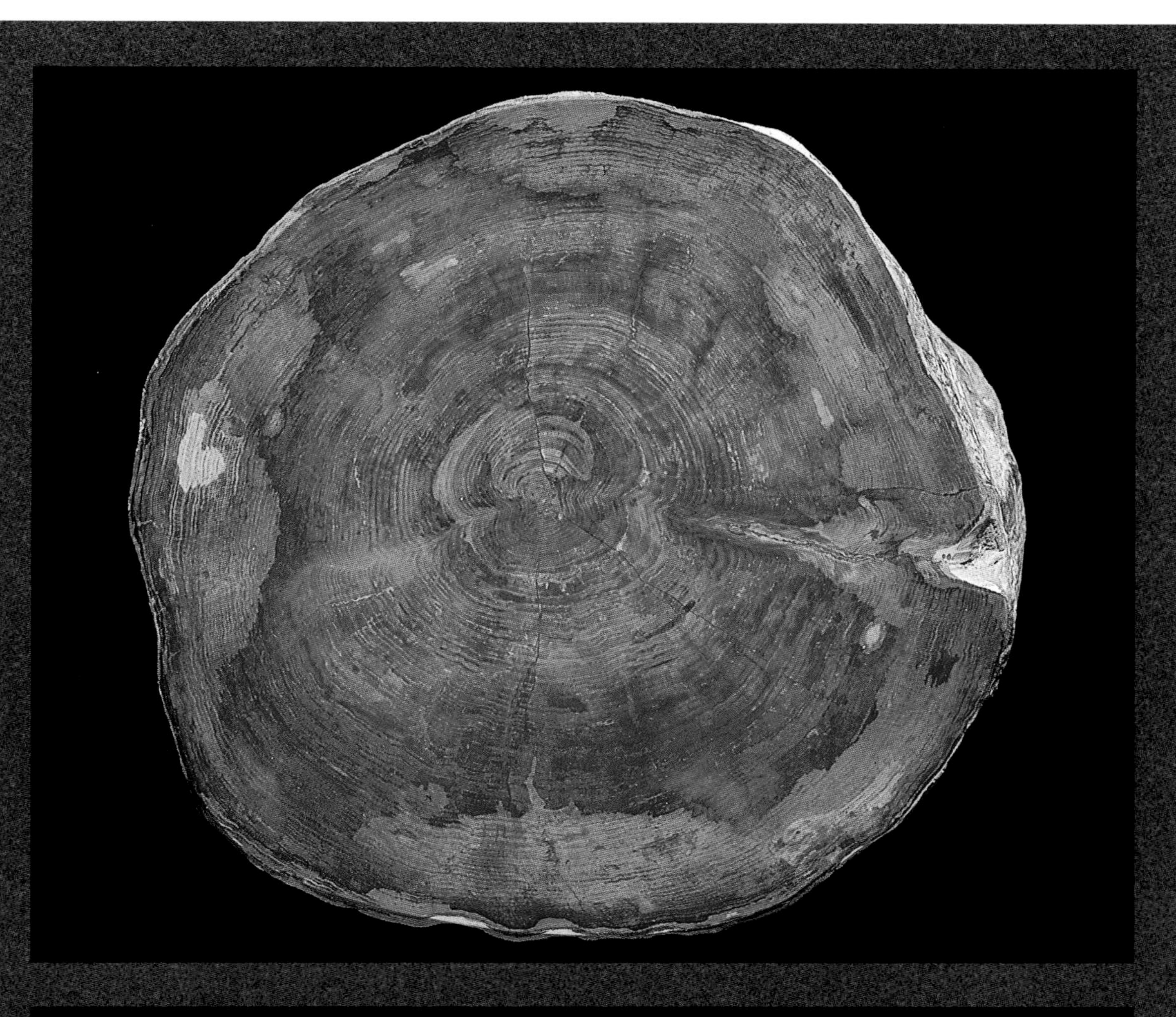

Australia
Mesozoic; Jurassic
conifer with mollusk bore tunnels
Daniels 19.5 cm

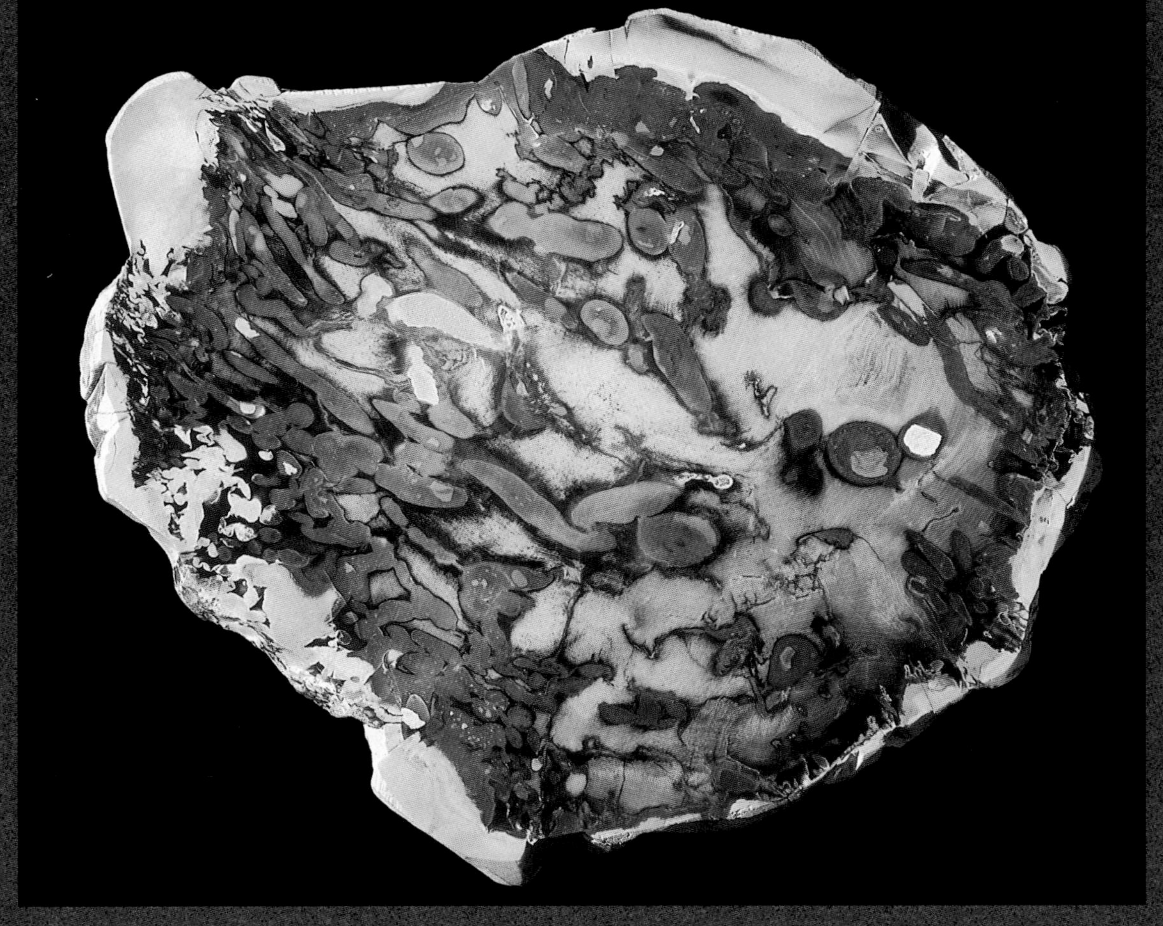

MADAGASCAR

Triassic age wood from Madagascar varies widely in quality. The best specimens are well silicified and have an excellent color range. These conifer specimens are predominantly *Araucaria* and are often found in full round logs.

Madagascar
Mesozoic;
Triassic
conifer
Daniels
12 cm

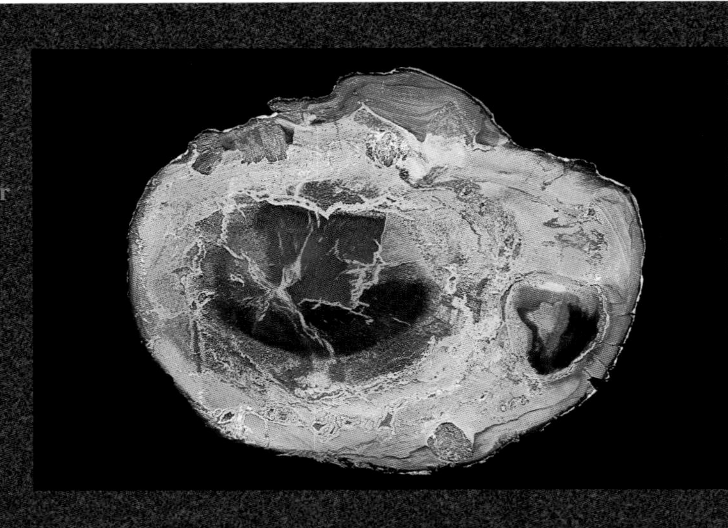

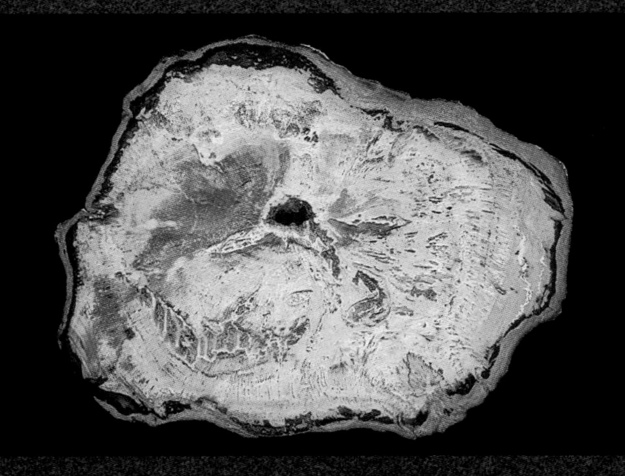

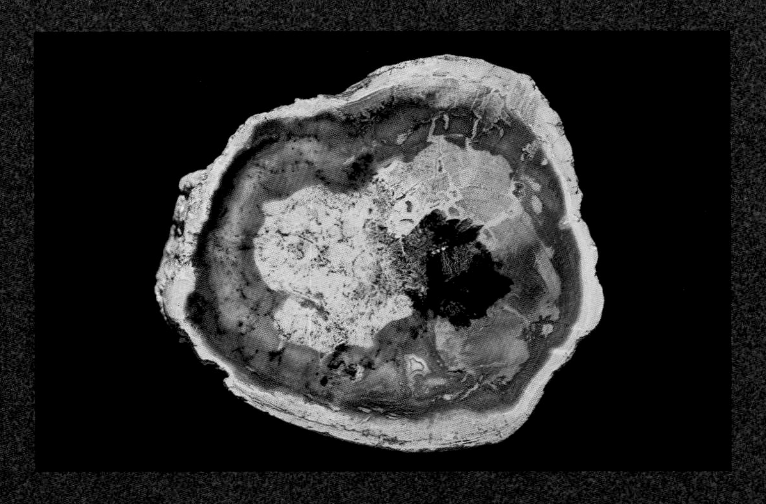

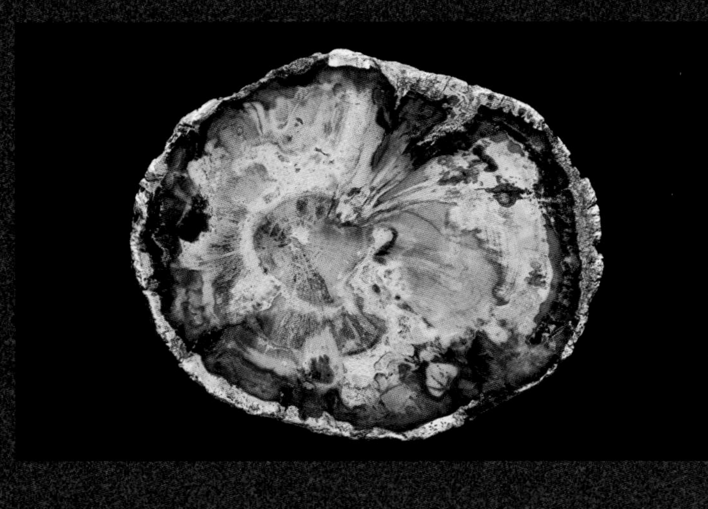

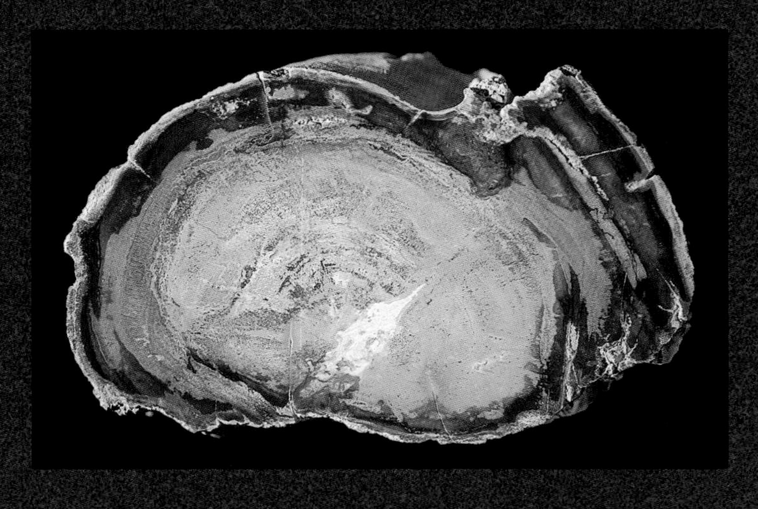

Madagascar
Mesozoic; Triassic
conifer
Daniels 12 cm

Madagascar
Mesozoic; Triassic
conifer
Daniels 5 cm

Madagascar
Mesozoic; Triassic
conifer
Daniels 5.5 cm

Madagascar
Mesozoic; Triassic
conifer
Daniels 9.5 cm

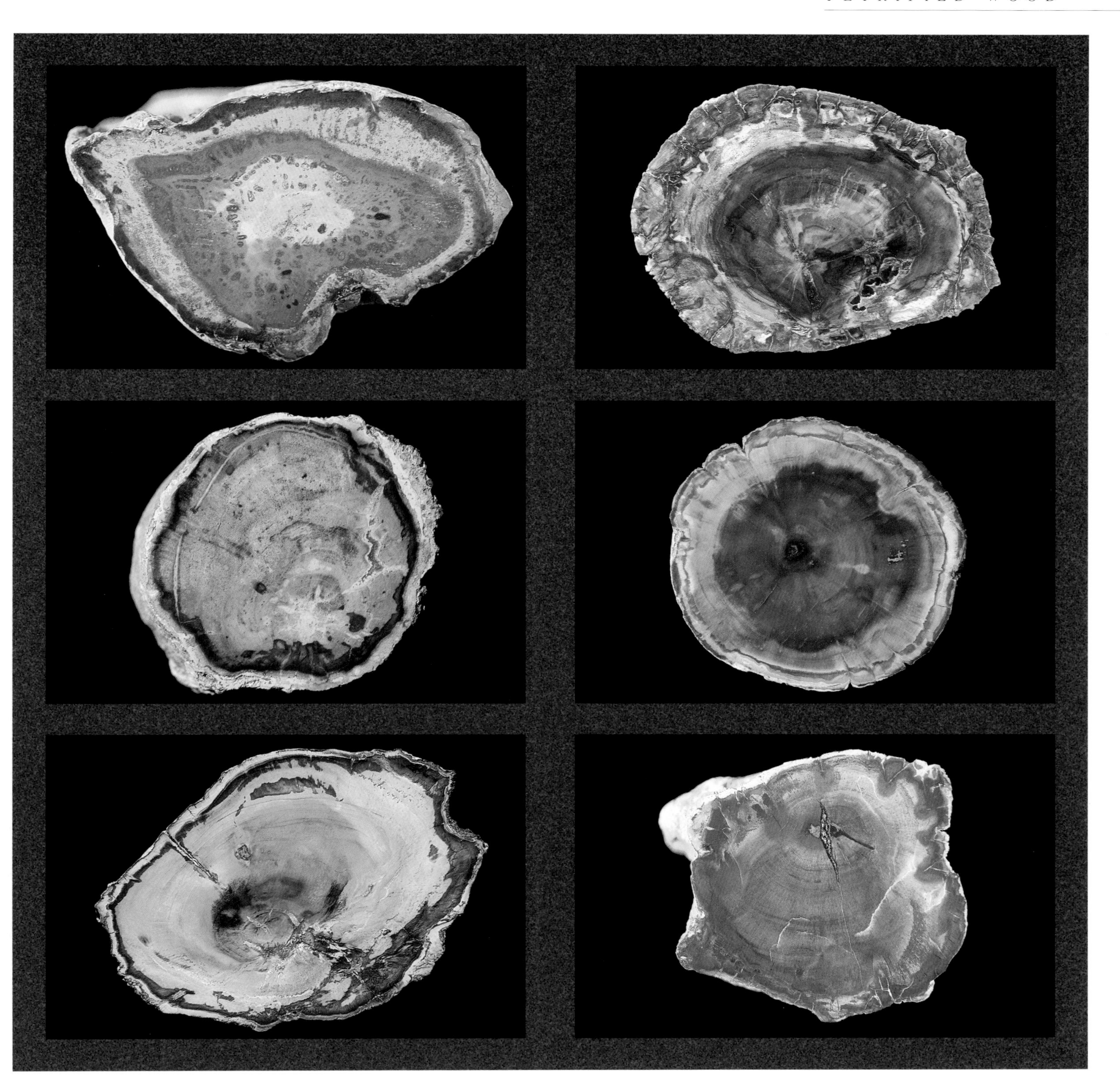

Madagascar
Mesozoic; Triassic
unidentified
Daniels 4 cm

Madagascar
Mesozoic; Triassic
conifer
Daniels 4.5 cm

Madagascar
Mesozoic; Triassic
conifer
Daniels 14 cm

Madagascar
Mesozoic; Triassic
conifer
Daniels 13 cm

Madagascar
Mesozoic; Triassic
conifer
Daniels 6 cm

Madagascar
Mesozoic; Triassic
conifer
Daniels 9 cm

TURKEY

Petrified wood from Turkey has recently become available to collectors in the West. Specimens from the area of Kizilcahaman in north-central Turkey are Tertiary in age and have been identified as juniper and cypress. These specimens have a white, chalky exterior and are very solid within, displaying well-preserved growth rings in an attractive range of colors from ivory to pink and lavender, and tan to brown.

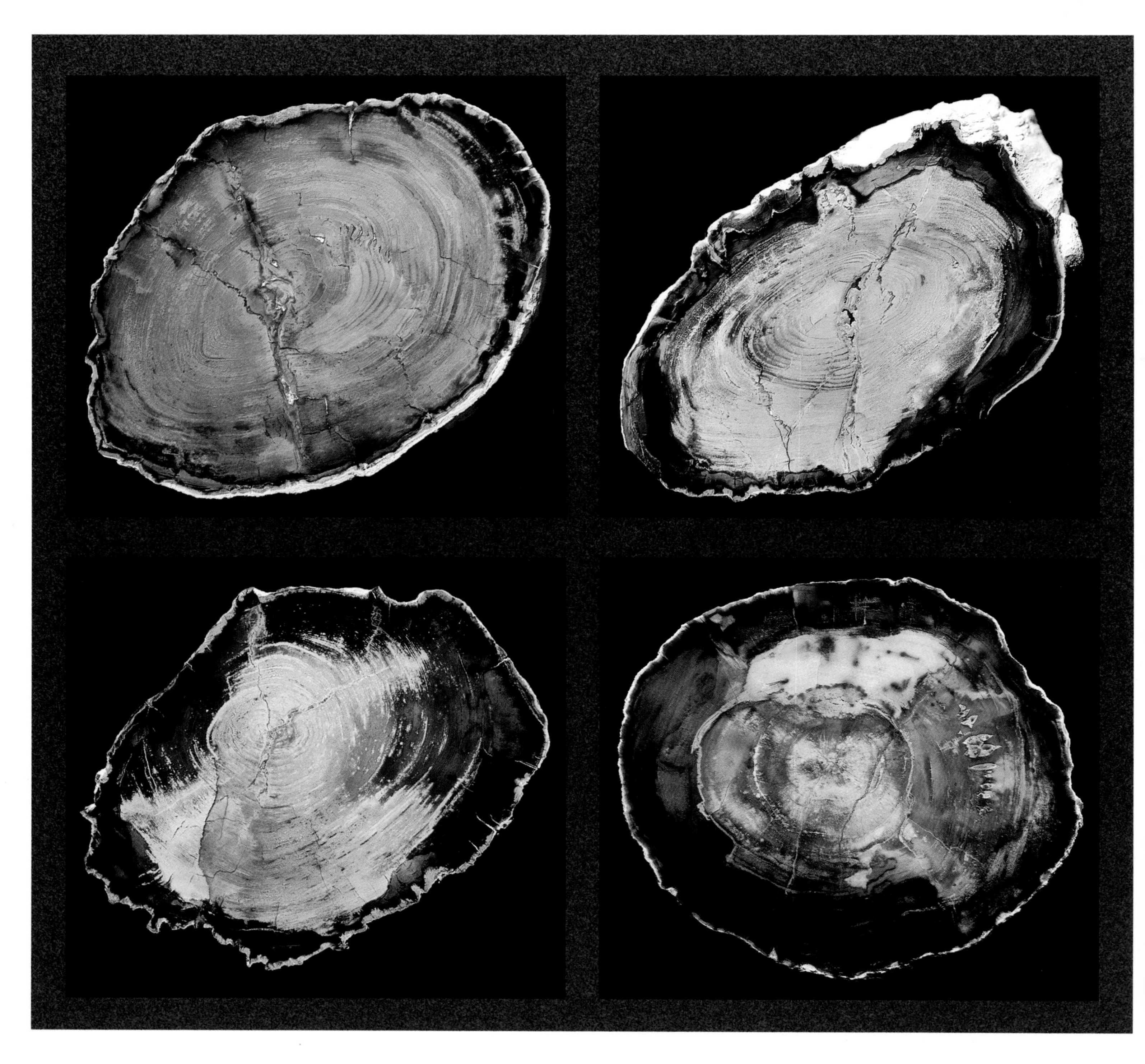

(UPPER LEFT)
Turkey [Kizilcahaman]
Cenozoic; Tertiary
***Juniperus* [Juniper]**
Cataldo 13 cm

(LOWER LEFT)
Turkey [Kizilcahaman]
Cenozoic; Tertiary
unidentified
Daniels 14 cm

(UPPER RIGHT)
Turkey [Kizilcahaman]
Cenozoic; Tertiary
unidentified
Daniels 15 cm

(LOWER RIGHT)
Turkey [Kizilcahaman]
Cenozoic; Tertiary
***Cupressus* [cypress]**
Daniels 13 cm

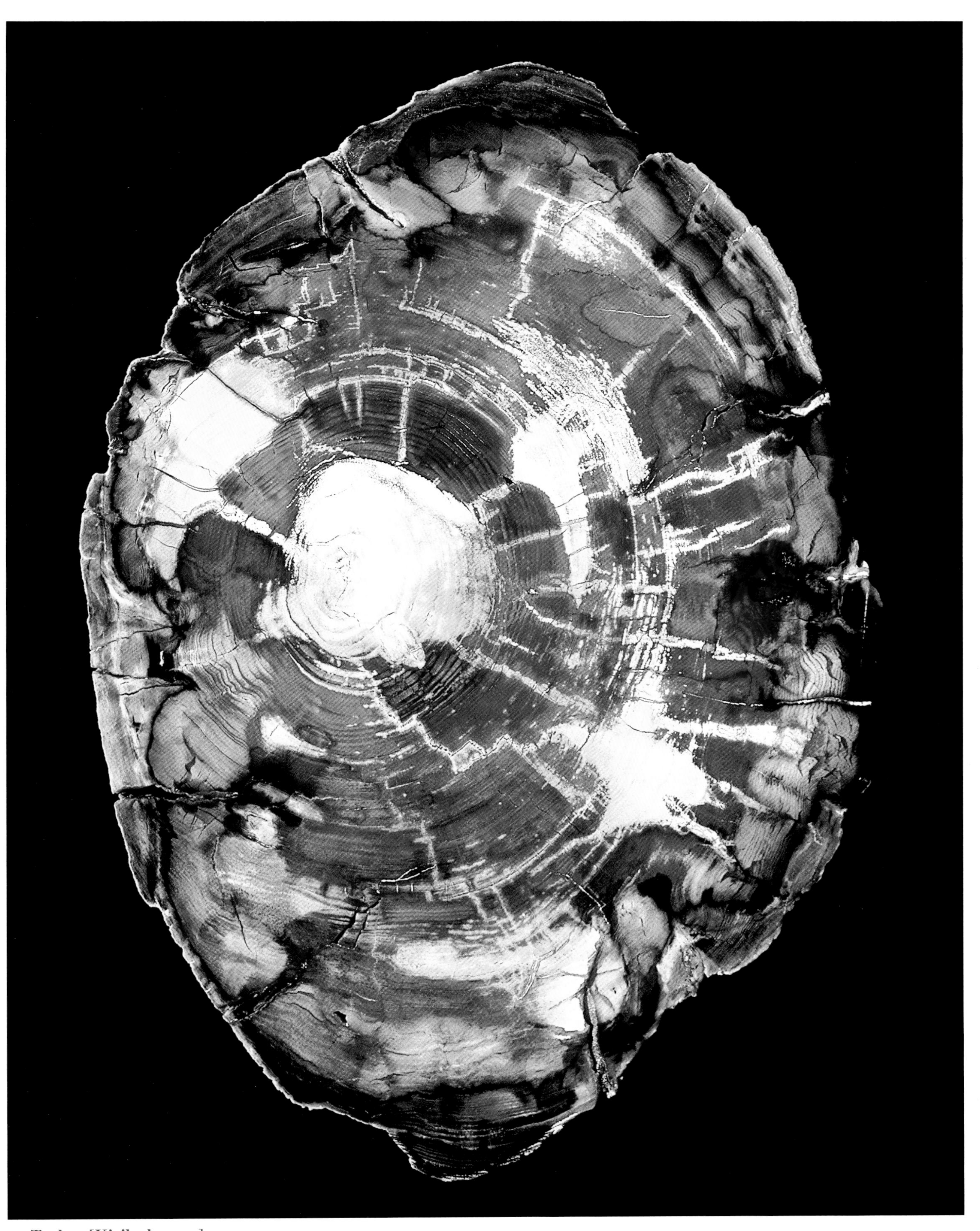

Turkey [Kizilcahaman]
Cenozoic; Tertiary
unidentified
Daniels 18 cm

ARIZONA

Arizona specimens are commonly from the Triassic Chinle Formation. Because of the quantity of materials available and the beauty, vibrance, and range of colors, specimens from this area reside in museums and collections worldwide. The majority of wood from this forest is *Araucarioxylon arizonicum*, but lesser quantities of *Woodworthia, Schilderia*, and other woods occur. Brightly colored *Araucarioxylon* from the area of Petrified Forest National Park is known as "rainbow wood." Colors vary widely, depending on the area and the member of the Chinle Formation in which the wood petrified.

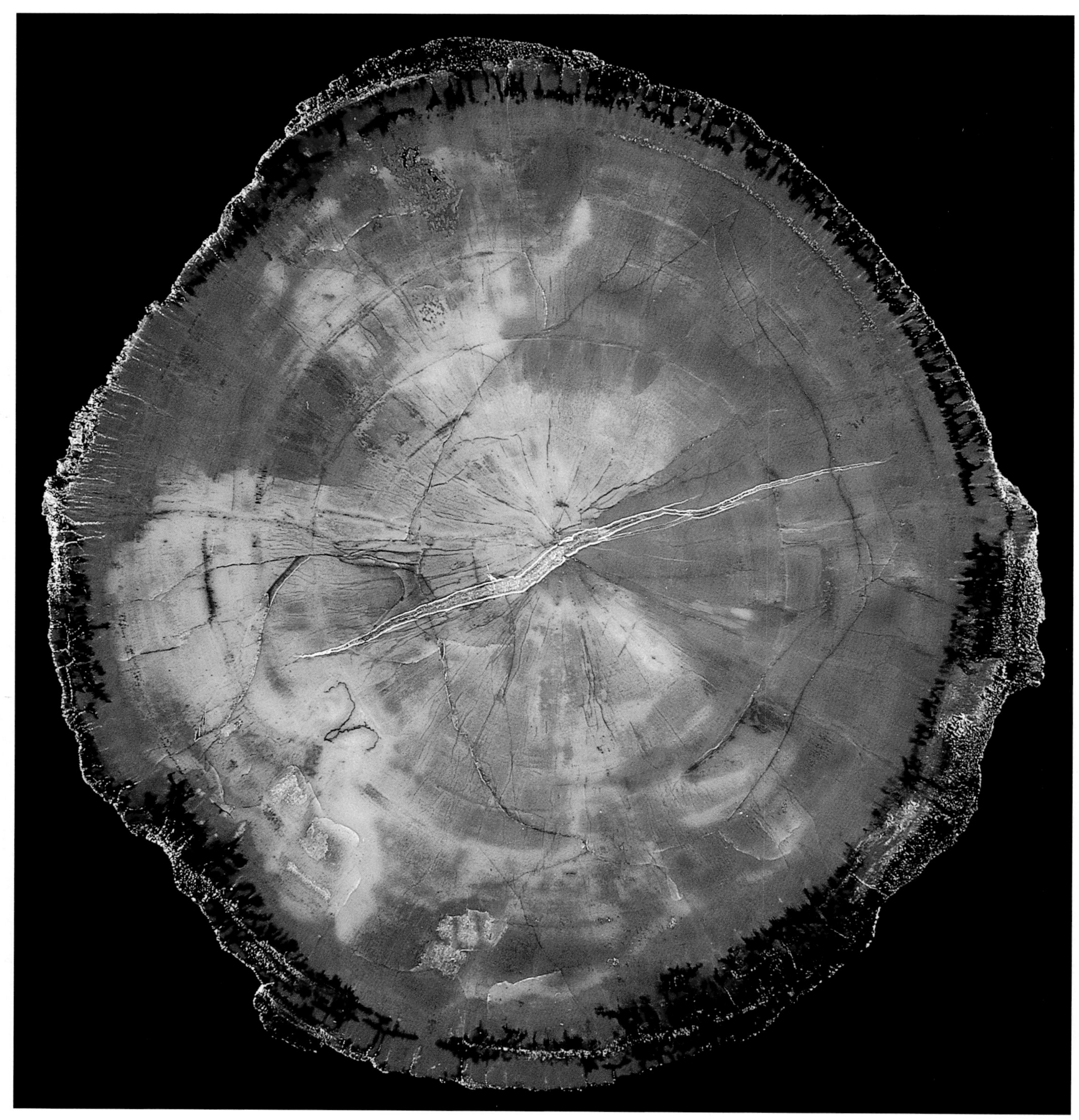

Arizona, USA
Mesozoic; Triassic
Araucarioxylon arizonicum
fm: Chinle
Jones 23.5 cm

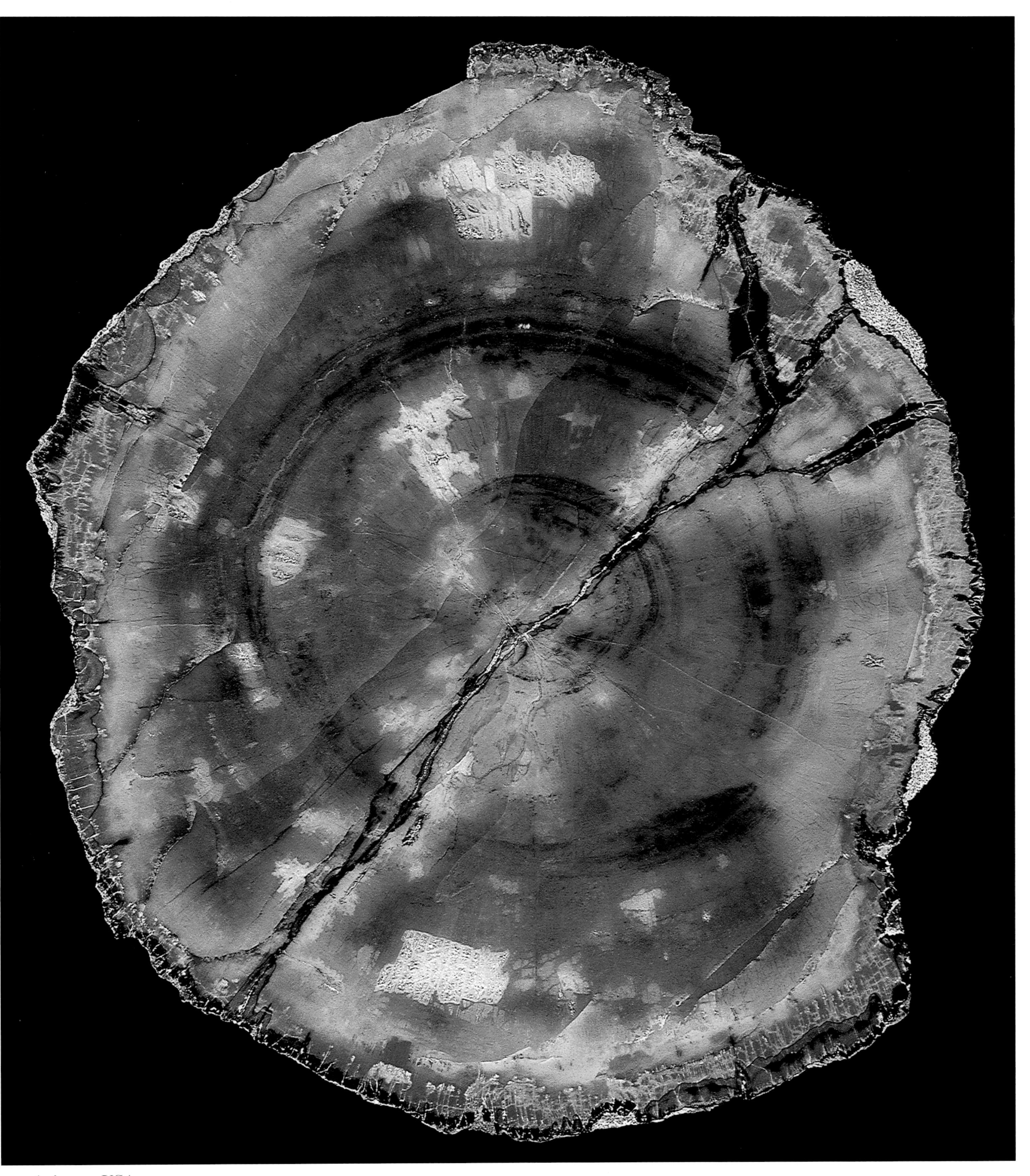

Arizona, USA
Mesozoic; Triassic
Araucarioxylon arizonicum **fm: Chinle**
Gray 21.5 cm

Arizona, USA
Mesozoic; Triassic
Dadoxylon fm: Chinle
Daniels 19.5 cm

Arizona, USA
Mesozoic; Triassic
unidentified fm: Chinle
Daniels 11 cm

Arizona, USA [Winslow]
Mesozoic; Triassic
conifer fm: Chinle
Gray 21 cm

Arizona, USA
Mesozoic; Triassic
conifer fm: Chinle
Gray 6 cm

Arizona, USA
Mesozoic; Triassic
Araucarioxylon arizonicum fm: Chinle
Gray 17 cm

Arizona, USA
Mesozoic; Triassic
Araucarioxylon arizonicum fm: Chinle
Gray 12 cm

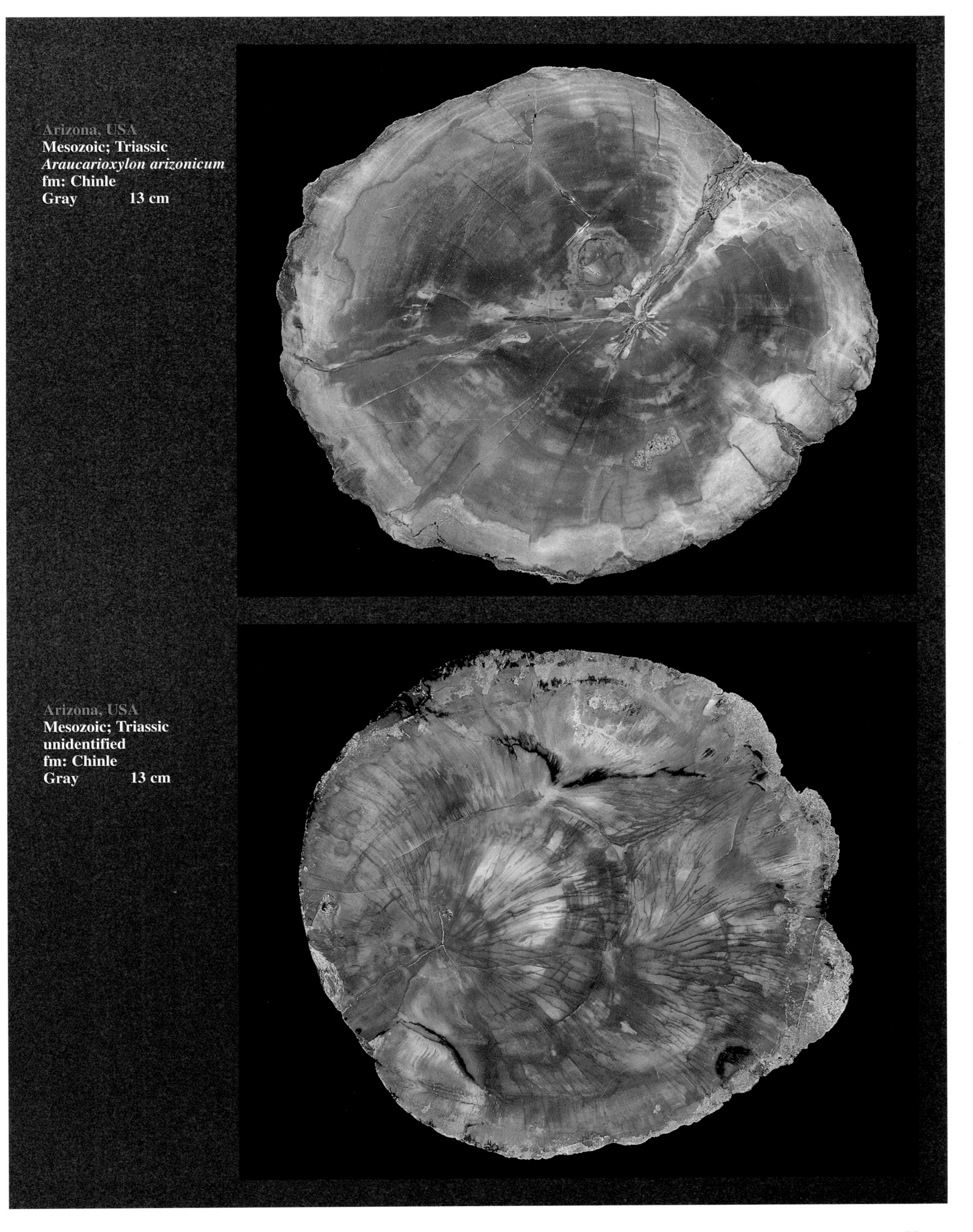

Arizona, USA
Mesozoic; Triassic
Araucarioxylon arizonicum
fm: Chinle
Gray 13 cm

Arizona, USA
Mesozoic; Triassic
unidentified
fm: Chinle
Gray 13 cm

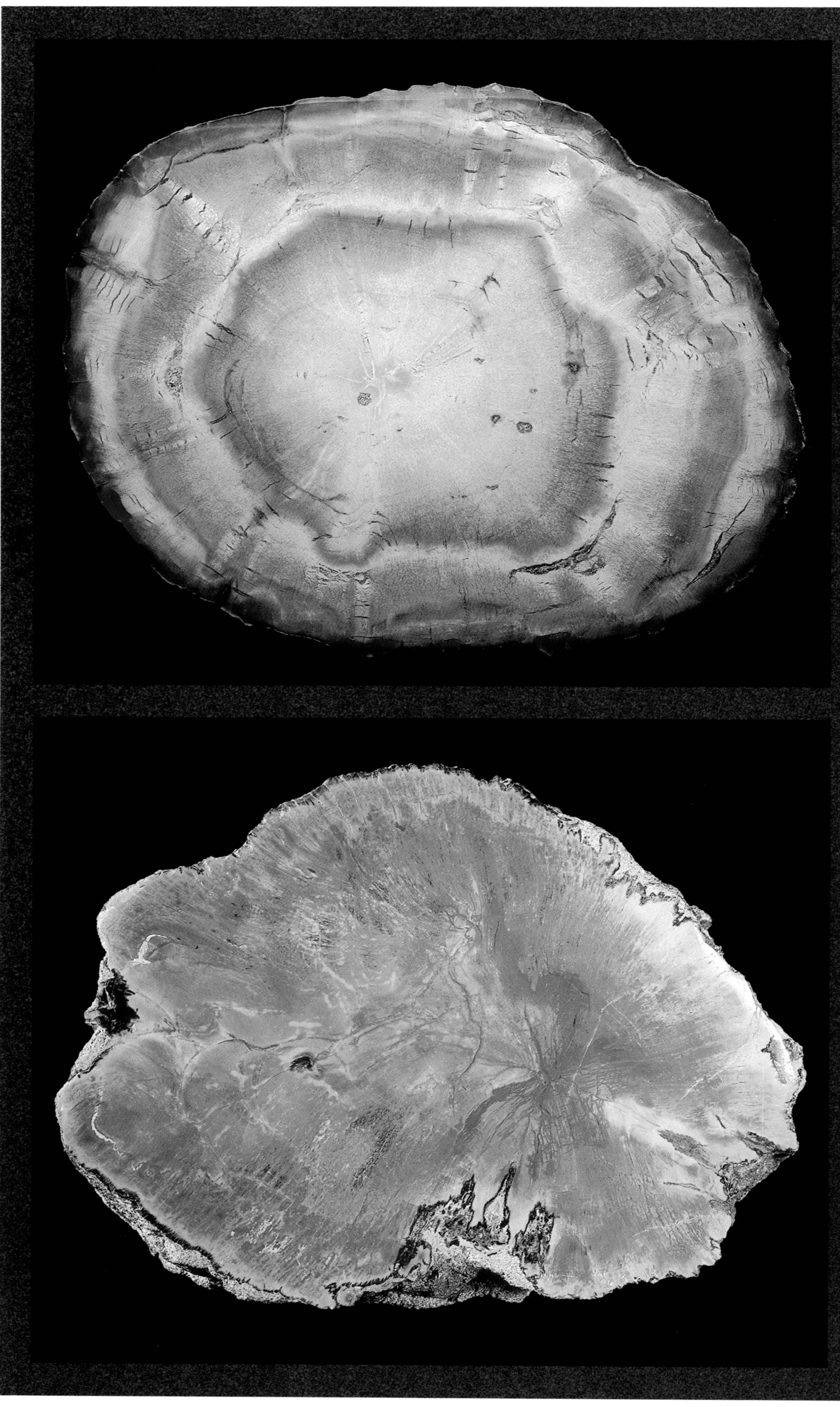

Arizona, USA
Mesozoic; Triassic
Woodworthia
fm: Chinle
Daniels 21 cm

Arizona, USA
Mesozoic; Triassic
unidentified
fm: Chinle
Daniels 16.5 cm

Arizona, USA
Mesozoic; Triassic
Araucarioxylon arizonicum **fm: Chinle**
Gray 14.5 cm

Arizona, USA
Mesozoic; Triassic
Araucarioxylon arizonicum
fm: Chinle
Jones 14 cm

Arizona, USA [St. Johns]
Mesozoic; Triassic
conifer fm: Chinle
Gray 21 cm

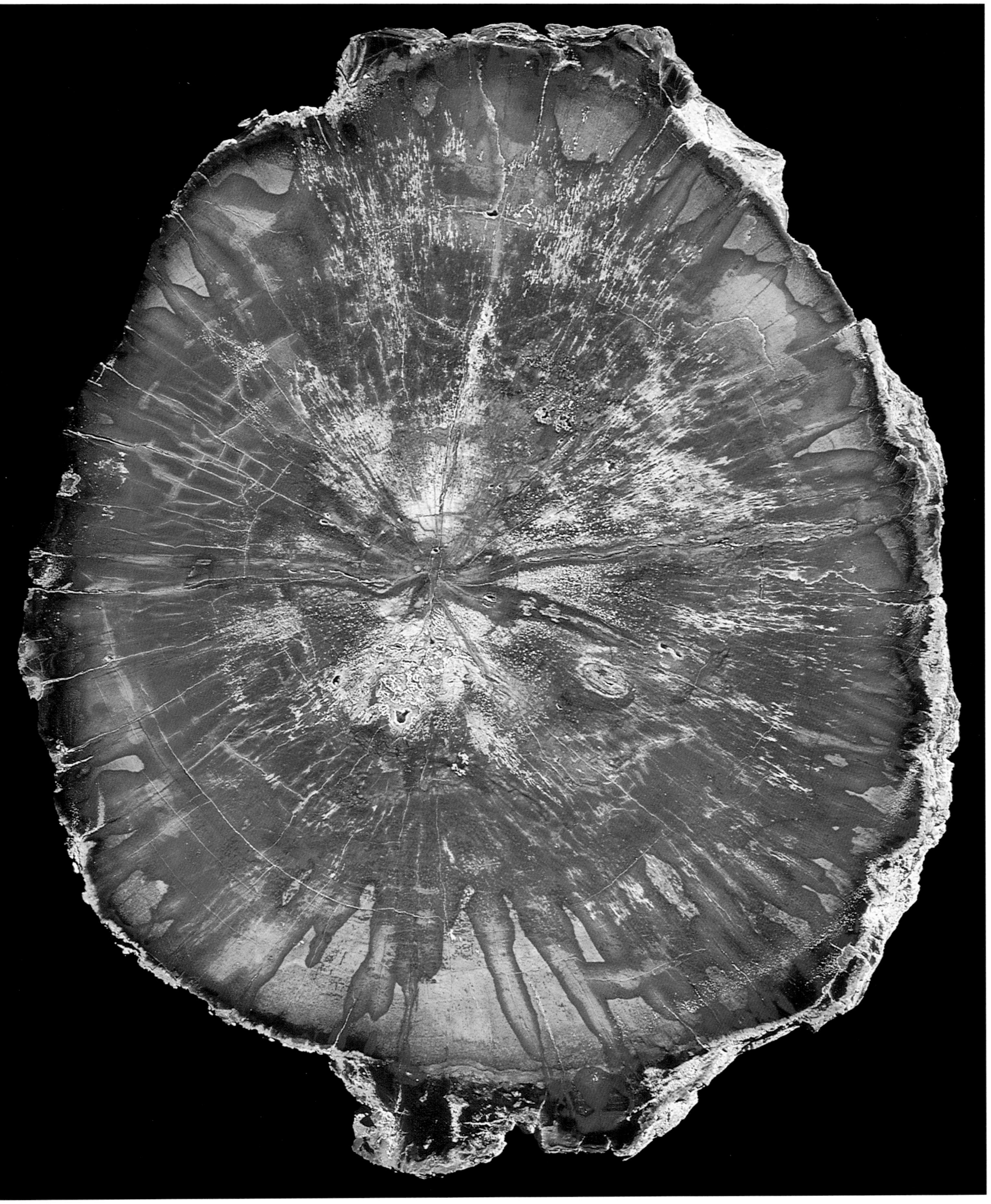

Arizona, USA	Mesozoic; Triassic	conifer
fm: Chinle	Daniels	36 cm

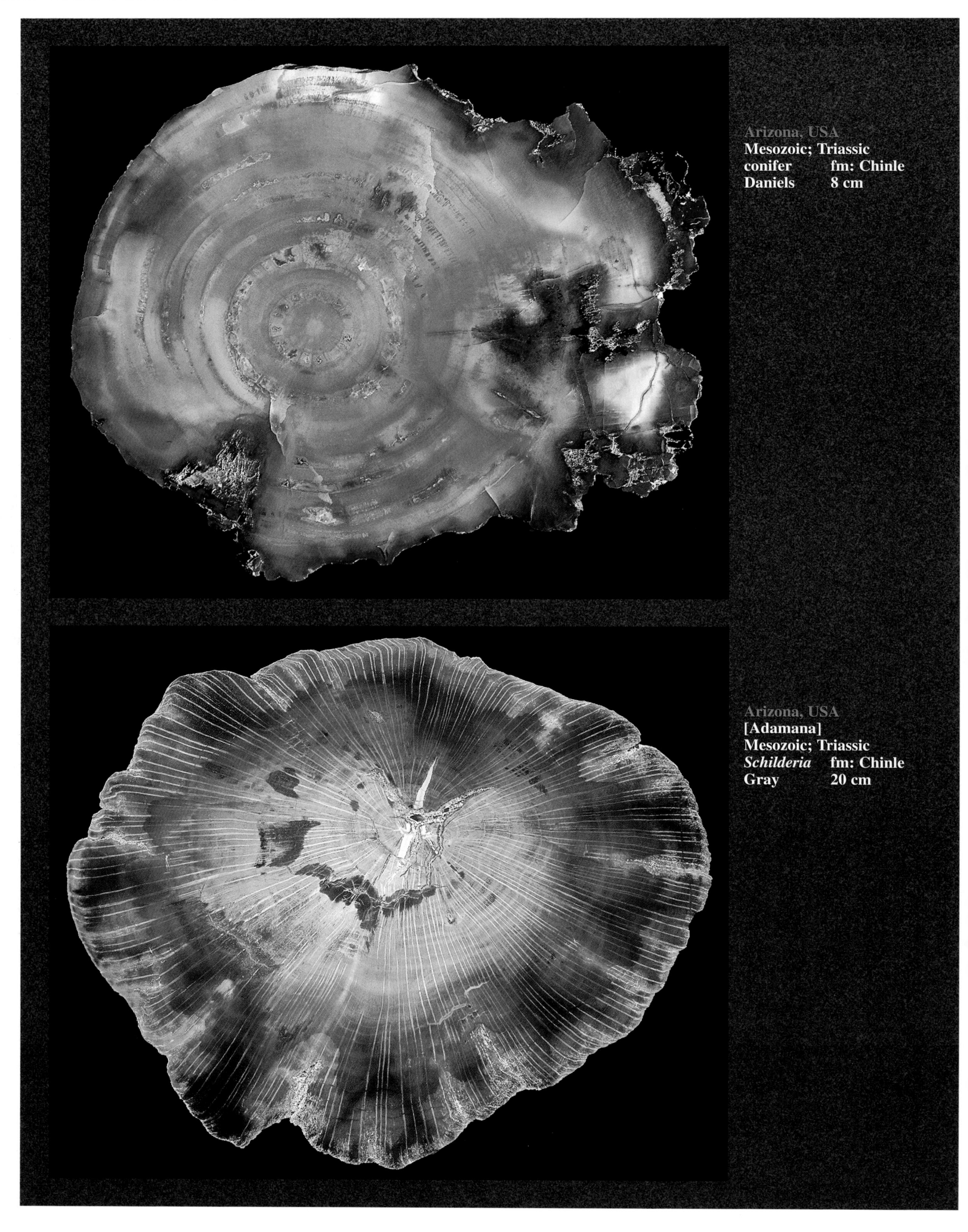

Arizona, USA
Mesozoic; Triassic
conifer fm: Chinle
Daniels 8 cm

Arizona, USA
[Adamana]
Mesozoic; Triassic
Schilderia fm: Chinle
Gray 20 cm

Arizona, USA [Sun Valley]	**Mesozoic; Triassic**	**conifer**
fm: Chinle	**Gray**	**28.5 cm**

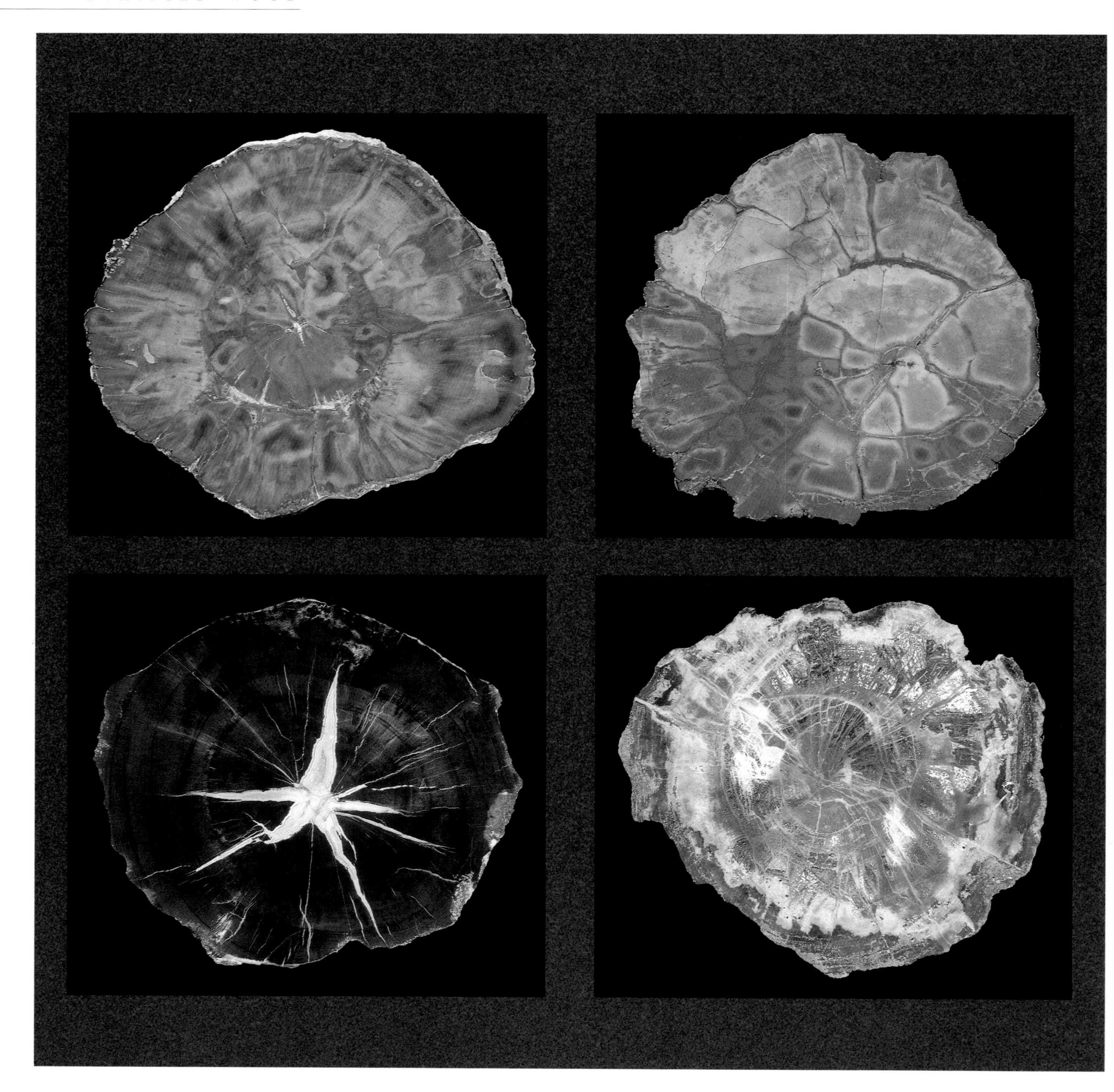

Arizona, USA
Mesozoic; Triassic
Araucarioxylon arizonicum **fm: Chinle**
Gray 14.5 cm

Arizona, USA
Mesozoic; Triassic
Araucarioxylon arizonicum **fm: Chinle**
Gray 15.5 cm

Arizona, USA [Adamana]
Mesozoic; Triassic
conifer fm: Chinle
Gray 14 cm

Arizona, USA
Mesozoic; Triassic
unidentified fm: Chinle
Gray 14 cm

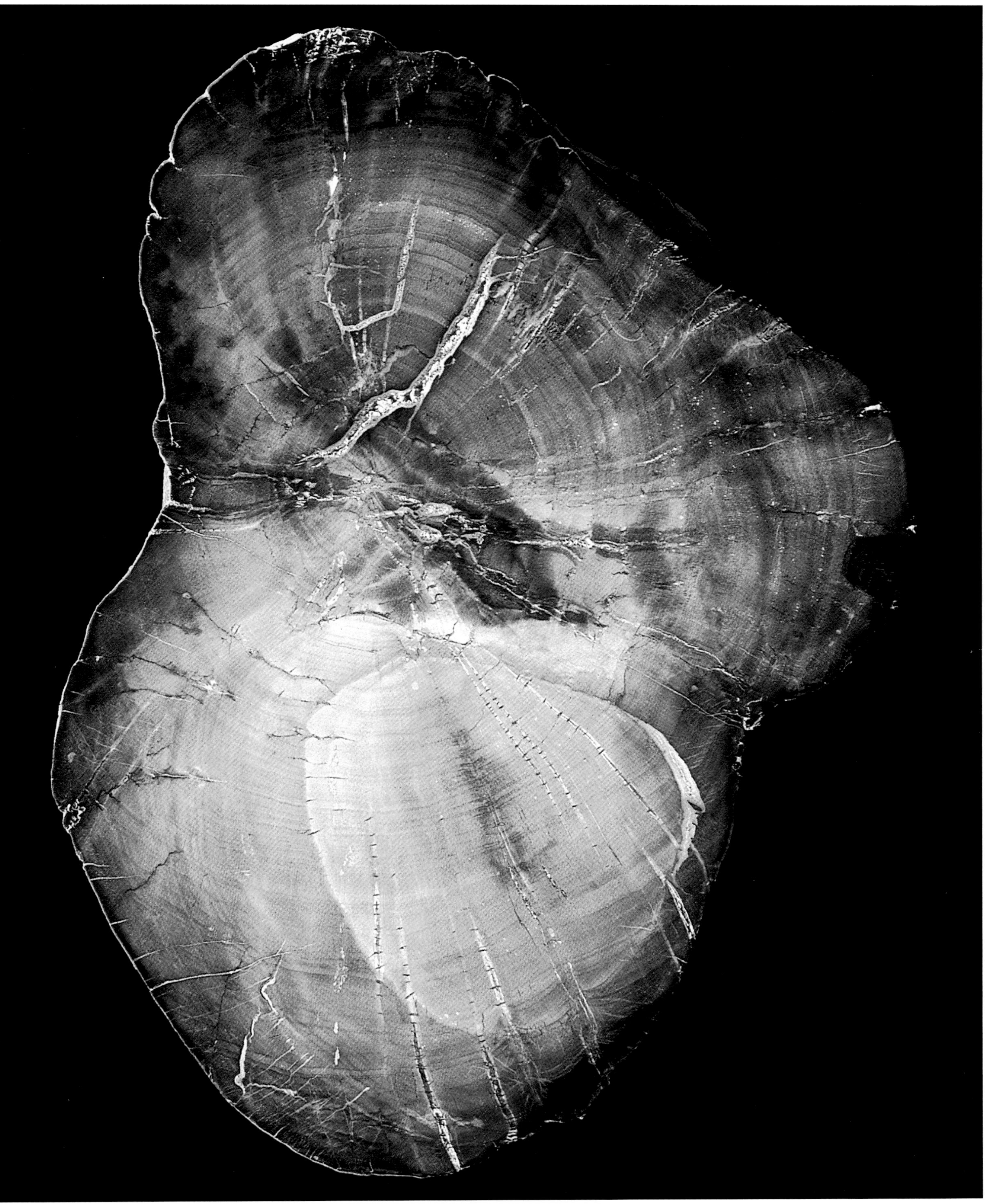

Arizona, USA	**Mesozoic; Triassic**	***Woodworthia***
fm: Chinle	**Daniels**	**31 cm**

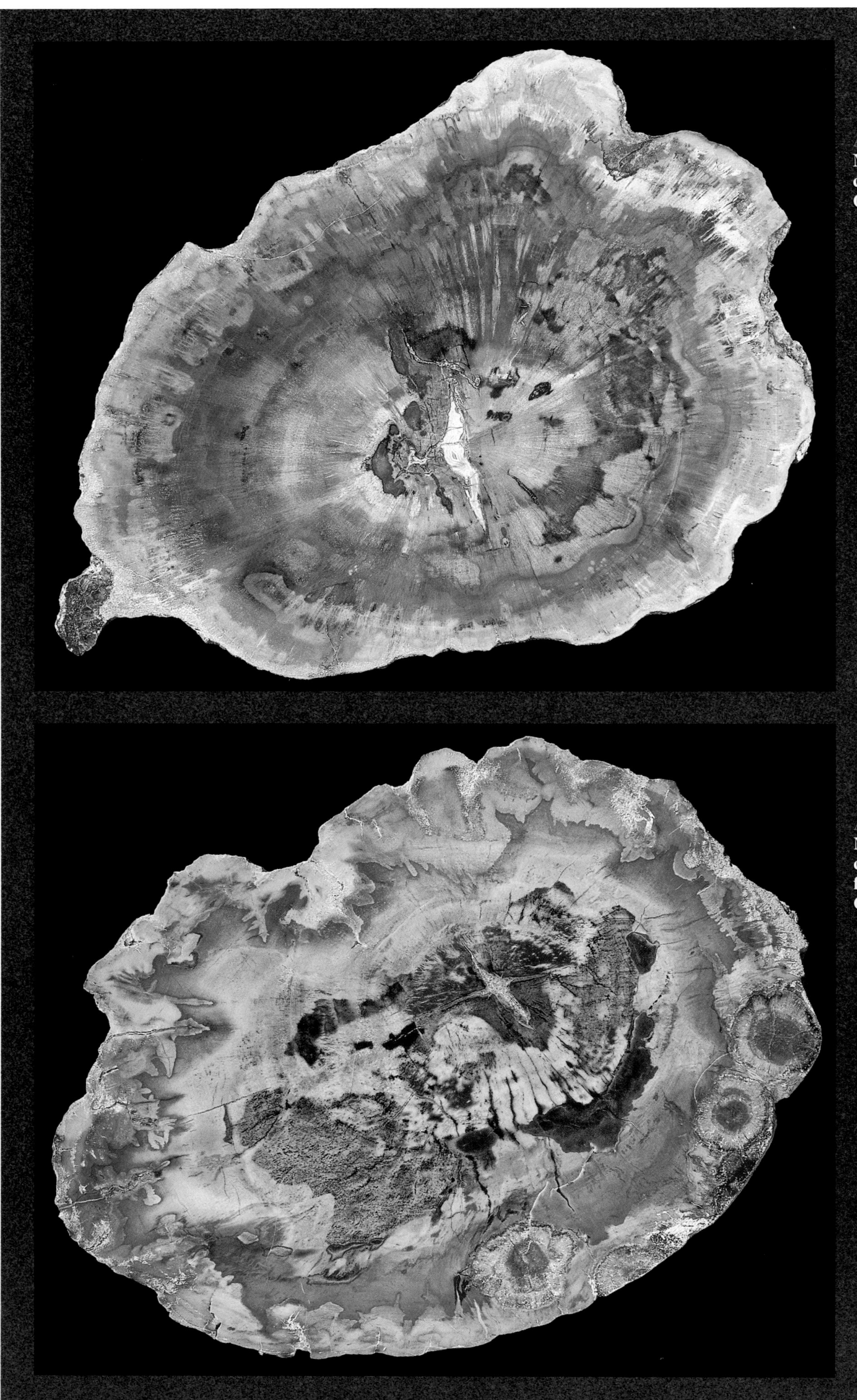

Arizona, USA [St. Johns]
Mesozoic; Triassic
conifer fm: Chinle
Gray 27 cm

Arizona, USA [St. Johns]
Mesozoic; Triassic
conifer with fungus
fm: Chinle
Gray 20 cm

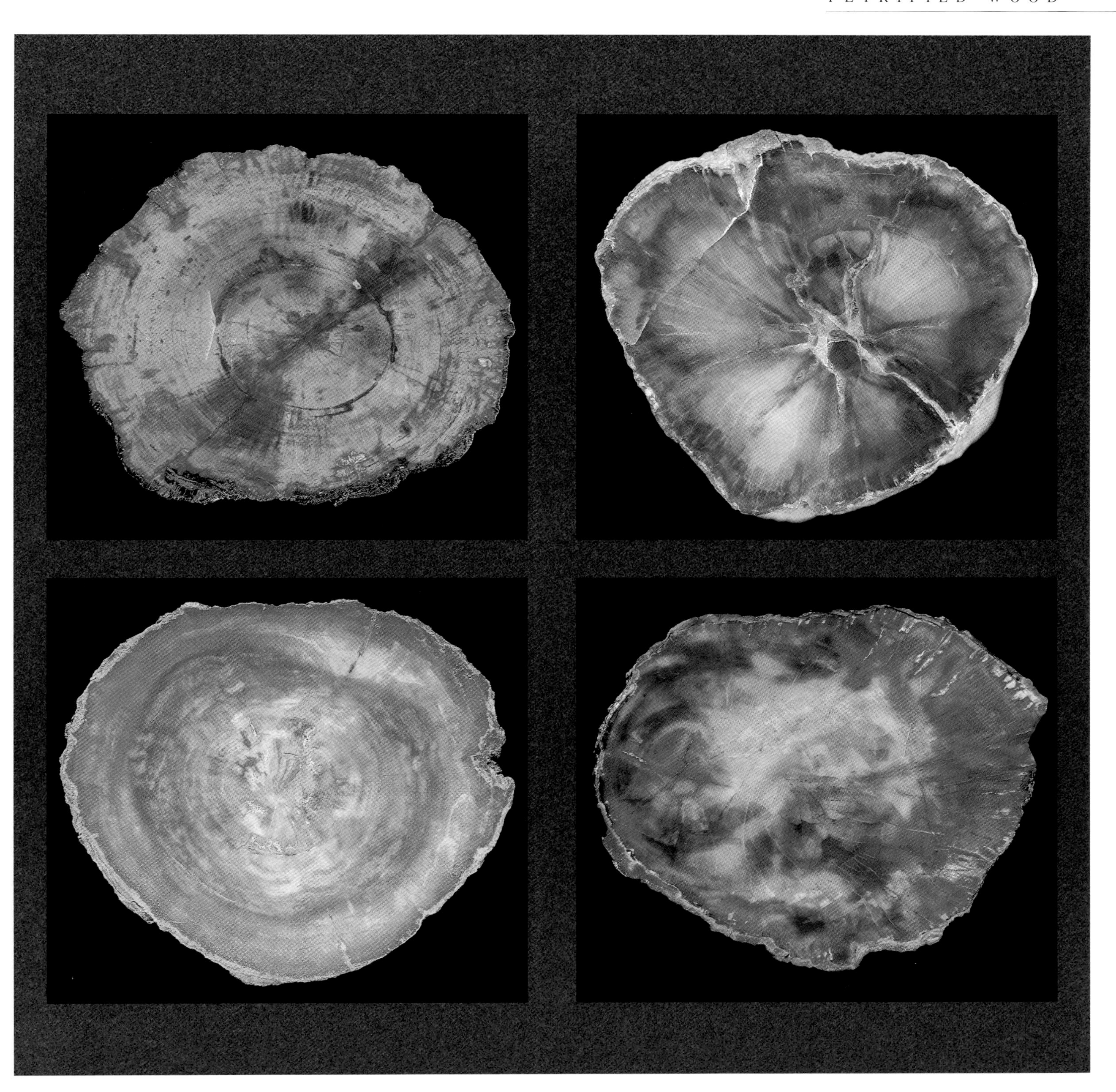

Arizona, USA
Mesozoic; Triassic
Araucarioxylon arizonicum **fm: Chinle**
Gray 18.5 cm

Arizona, USA
Mesozoic; Triassic
unidentified fm: Chinle
Gray 8 cm

Arizona, USA
Mesozoic; Triassic
Araucarioxylon arizonicum **fm: Chinle**
Jones 19.5 cm

Arizona, USA
Mesozoic; Triassic
conifer fm: Chinle
Justman 16.5 cm

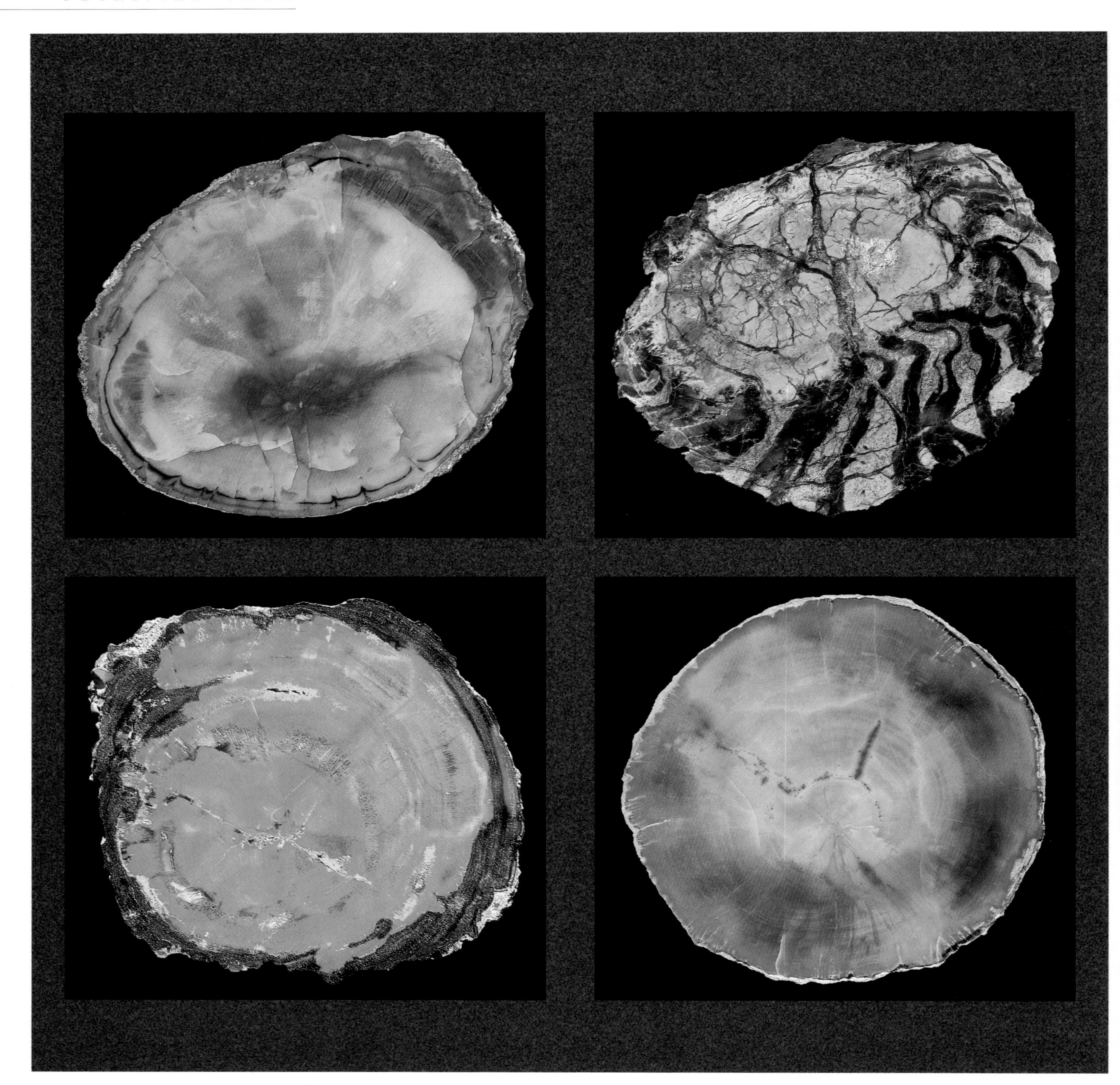

Arizona, USA
Mesozoic; Triassic
Araucarioxylon arizonicum **fm: Chinle**
Gray 13.5 cm

Arizona, USA
Mesozoic; Triassic
unidentified fm: Chinle
Gray 11 cm

Arizona, USA
Mesozoic; Triassic
Araucarioxylon arizonicum **fm: Chinle**
Daniels 25.5 cm

Arizona, USA
Mesozoic; Triassic
Araucarioxylon arizonicum **fm: Chinle**
Gray 9 cm

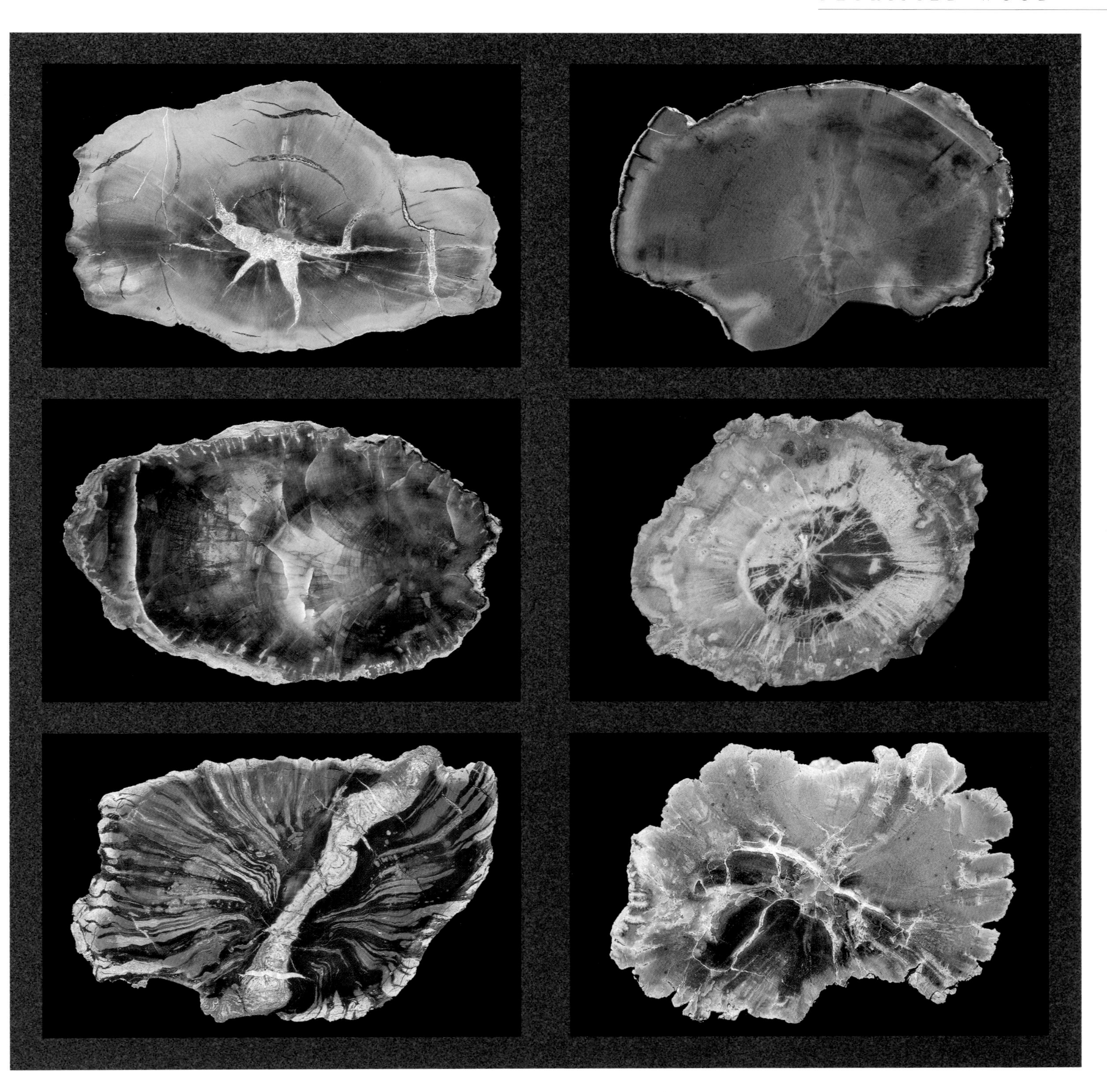

Arizona, USA [Winslow]
Mesozoic; Triassic
Woodworthia **fm: Chinle**
Gray 14.5 cm

Arizona, USA
Mesozoic; Triassic
conifer fm: Chinle
Daniels 15 cm

Arizona, USA
Mesozoic; Triassic
unidentified fm: Chinle
Scott 22 cm

Arizona, USA
Mesozoic; Triassic
Araucarioxylon arizonicum **fm: Chinle**
Daniels 8 cm

Arizona, USA
Mesozoic; Triassic
conifer fm: Chinle
Gray 10 cm

Arizona, USA [Winslow]
Mesozoic; Triassic
conifer fm: Chinle
Jones 7 cm

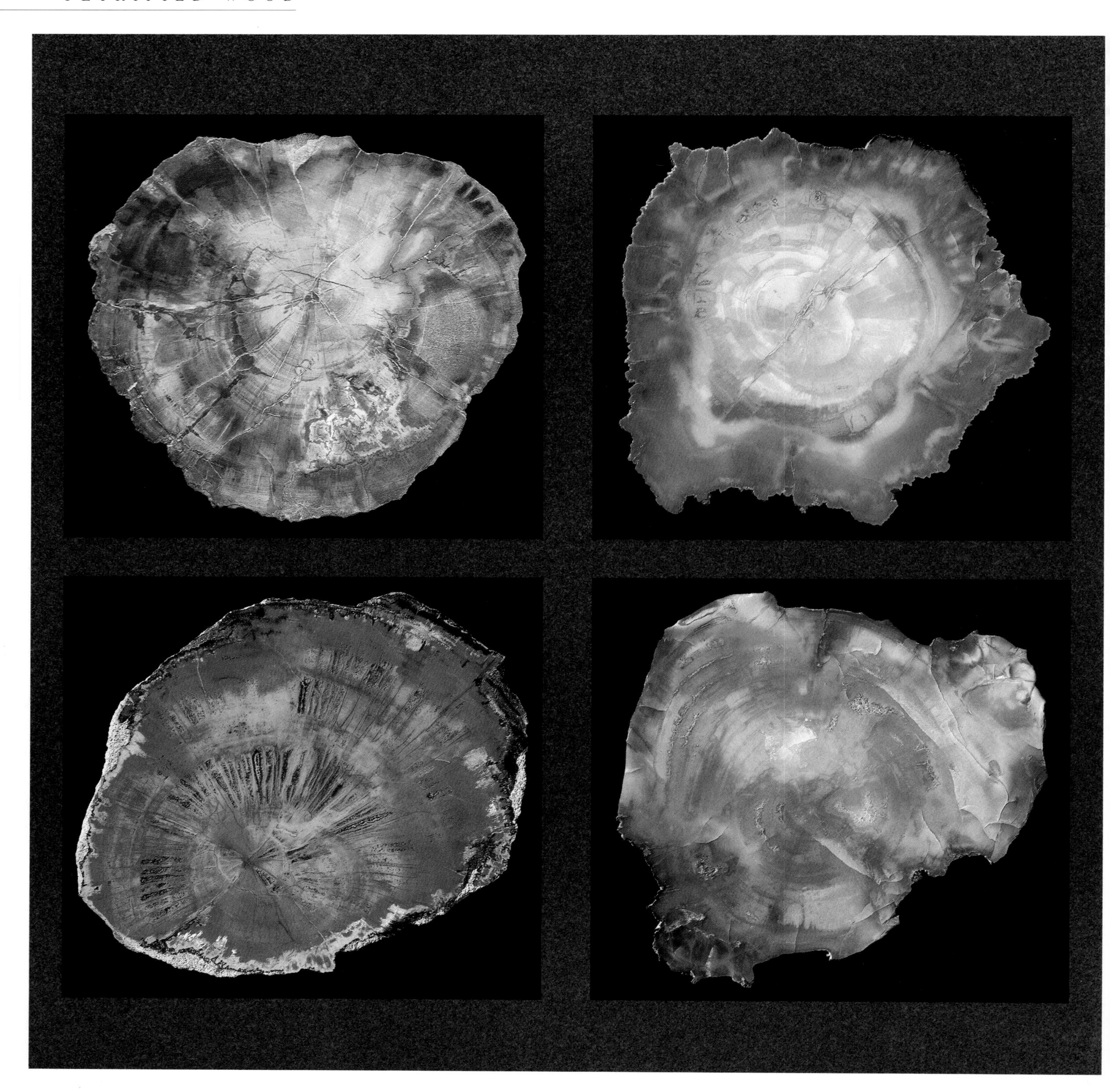

Arizona, USA
Mesozoic; Triassic
Araucarioxylon arizonicum **fm: Chinle**
Daniels 14 cm

Arizona, USA
Mesozoic; Triassic
conifer fm: Chinle
Daniels 12 cm

Arizona, USA
Mesozoic; Triassic
Araucarioxylon arizonicum **fm: Chinle**
Gray 17 cm

Arizona, USA
Mesozoic; Triassic
conifer fm: Chinle
Daniels 11 cm

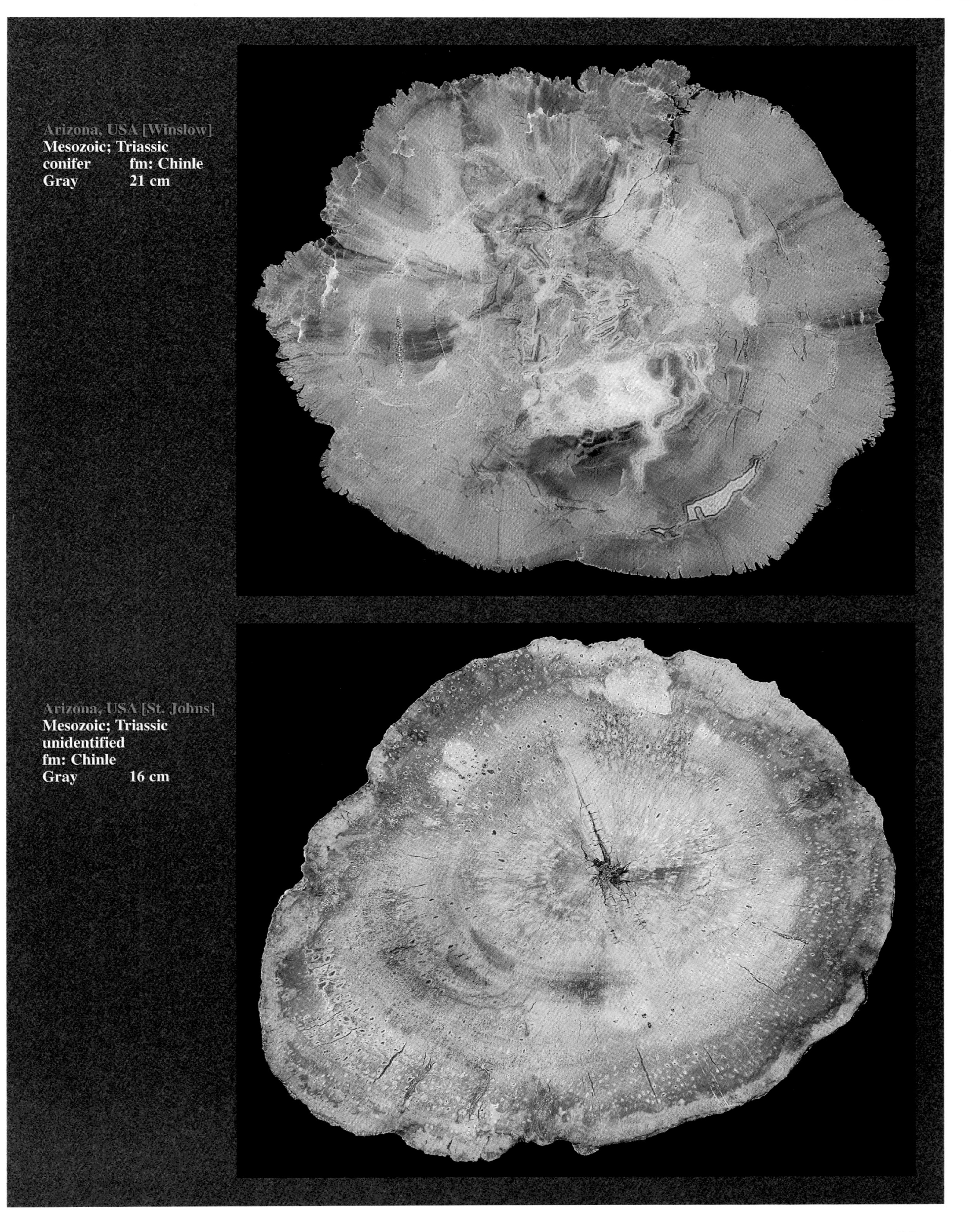

Arizona, USA [Winslow]
Mesozoic; Triassic
conifer fm: Chinle
Gray 21 cm

Arizona, USA [St. Johns]
Mesozoic; Triassic
unidentified
fm: Chinle
Gray 16 cm

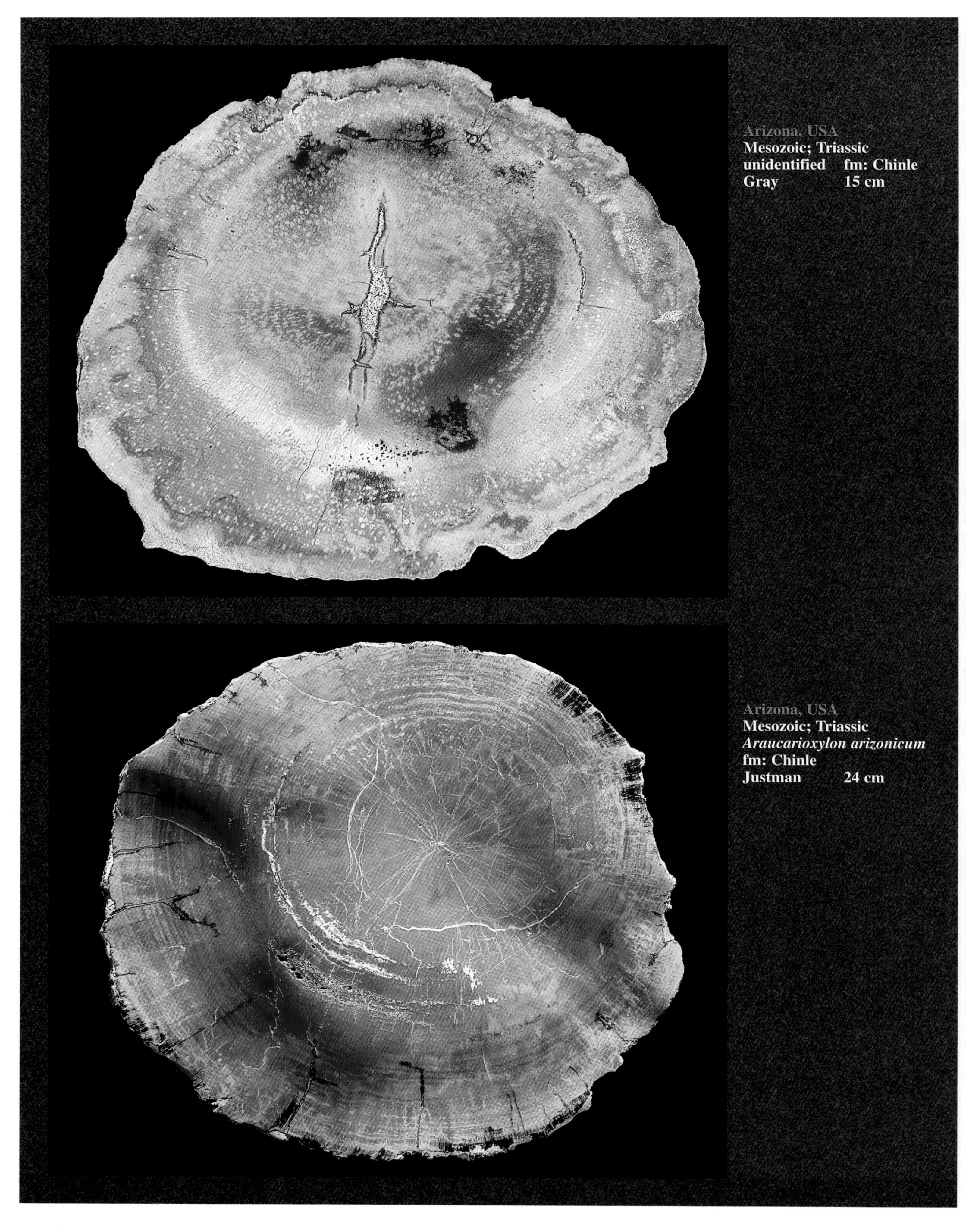

Arizona, USA
Mesozoic; Triassic
unidentified fm: Chinle
Gray 15 cm

Arizona, USA
Mesozoic; Triassic
Araucarioxylon arizonicum
fm: Chinle
Justman 24 cm

Arizona, USA
Mesozoic; Triassic
Araucarioxylon arizonicum
fm: Chinle
Daniels 15 cm

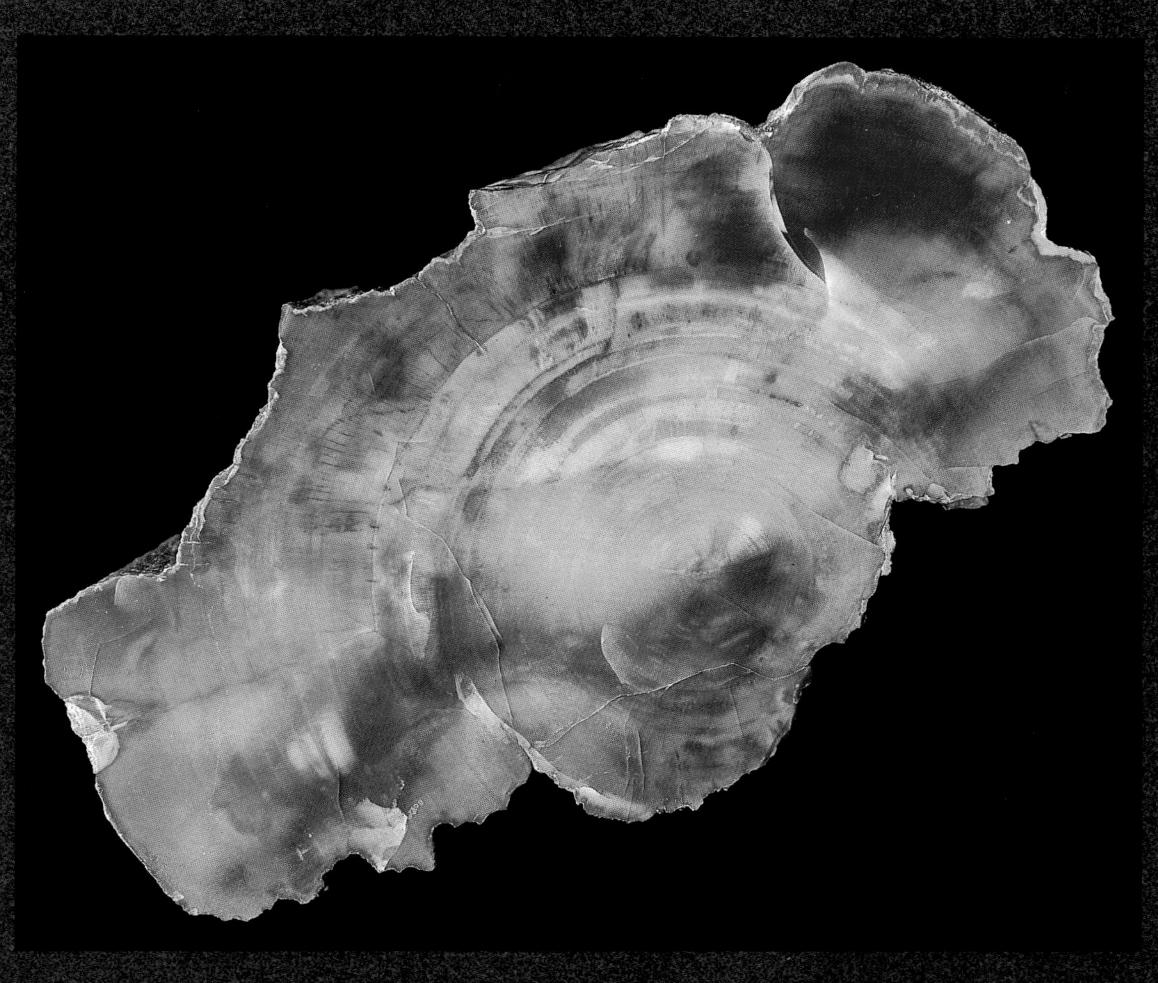

Arizona, USA
Mesozoic; Triassic
Araucarioxylon arizonicum
fm: Chinle
Gray 14 cm

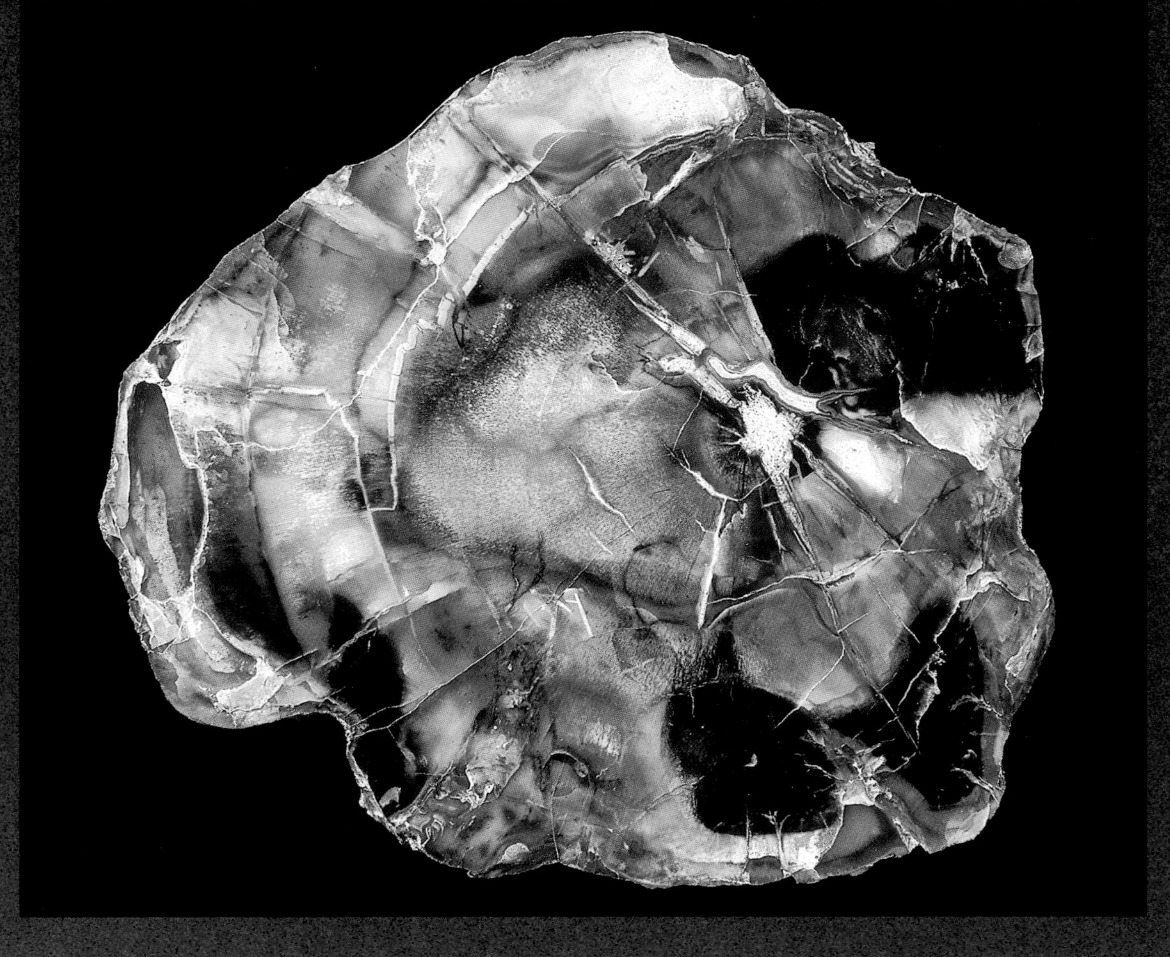

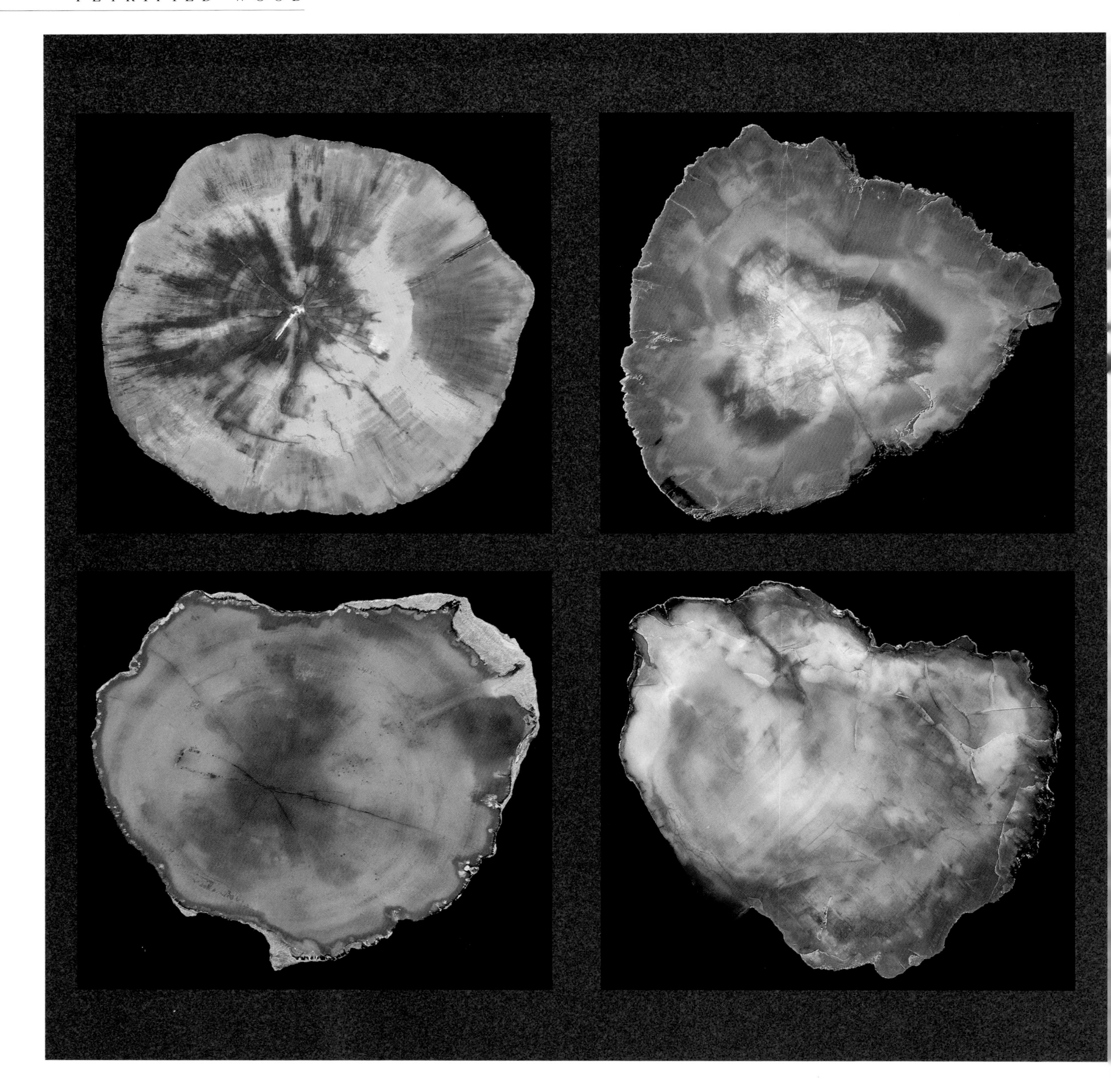

Arizona, USA
Mesozoic; Triassic
Araucarioxylon arizonicum **fm: Chinle**
Gray 15.5 cm

Arizona, USA
Mesozoic; Triassic
Araucarioxylon arizonicum **fm: Chinle**
Daniels 10 cm

Arizona, USA
Mesozoic; Triassic
Araucarioxylon arizonicum **fm: Chinle**
Gray 8 cm

Arizona, USA
Mesozoic; Triassic
Araucarioxylon arizonicum
fm: Chinle
Daniels 12.5 cm

Arizona, USA
Mesozoic; Triassic
Araucarioxylon arizonicum **fm: Chinle**
Daniels 13.5 cm

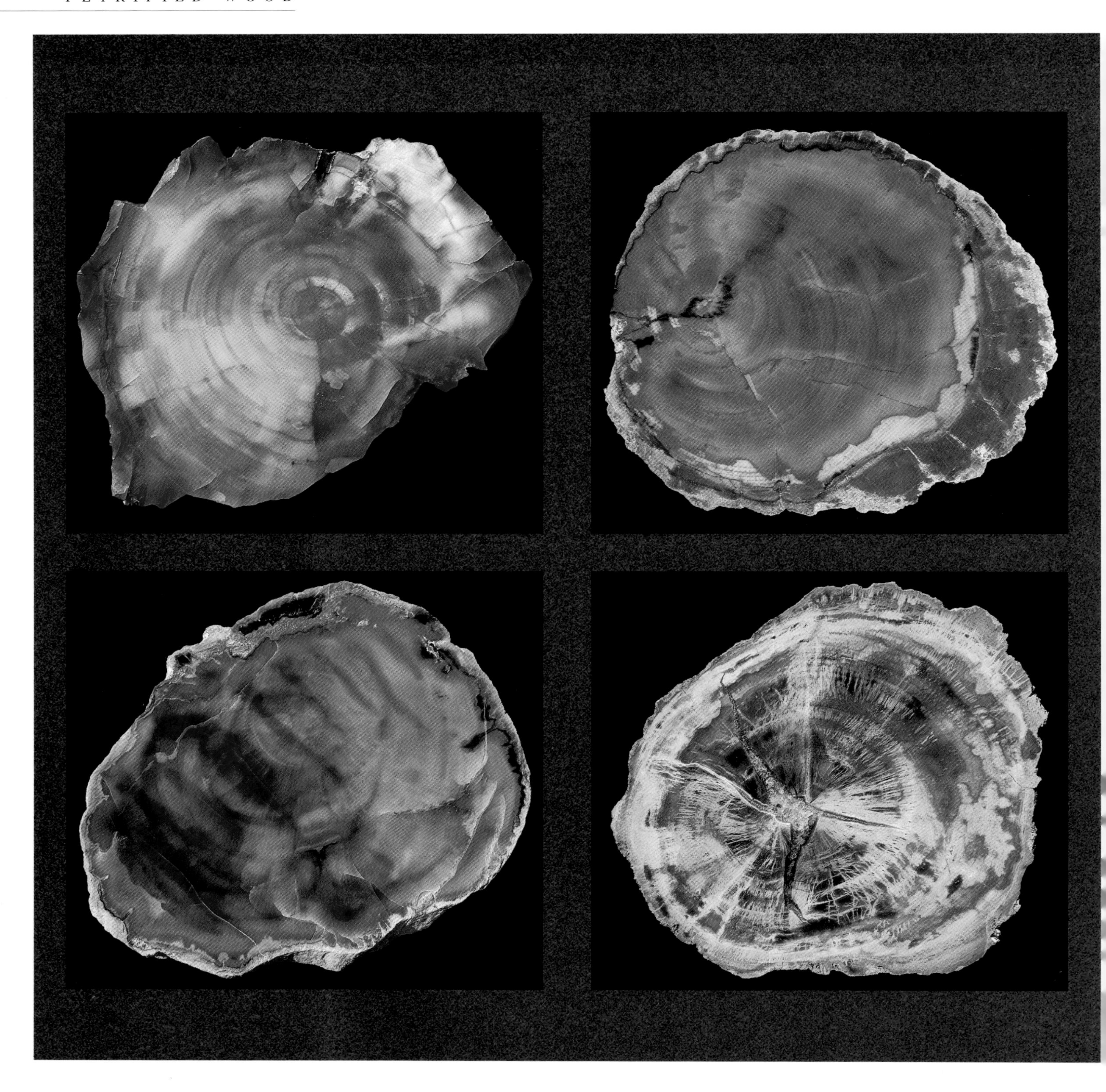

Arizona, USA
Mesozoic; Triassic
conifer fm: Chinle
Daniels 7.5 cm

Arizona, USA
Mesozoic; Triassic
Araucarioxylon arizonicum **fm: Chinle**
Gray 7.5 cm

Arizona, USA
Mesozoic; Triassic
Araucarioxylon arizonicum **fm: Chinle**
Gray 14.5 cm

Arizona, USA [St. Johns]
Mesozoic; Triassic
conifer fm: Chinle
Daniels 19.5 cm

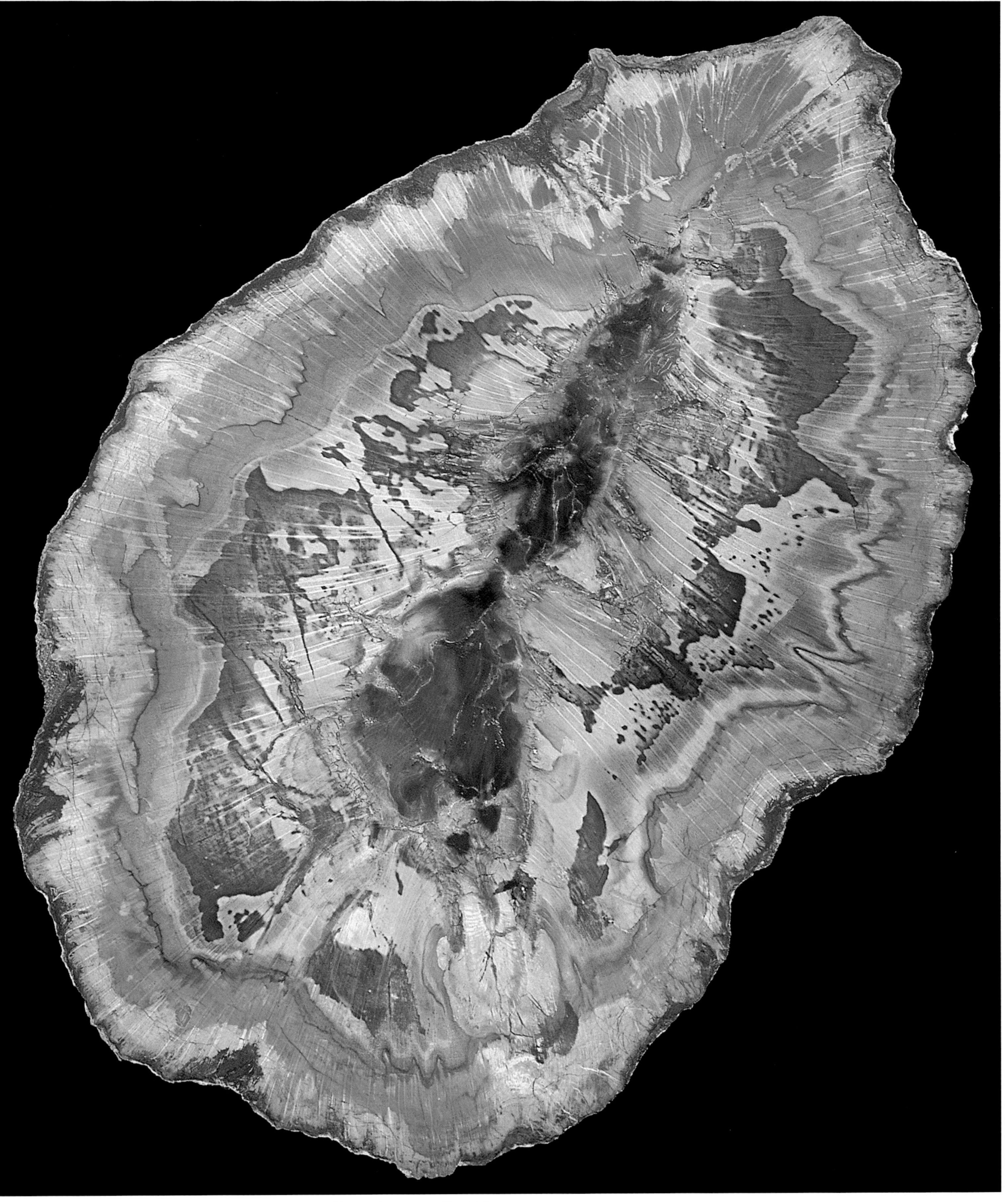

Arizona, USA [St. Johns]	**Mesozoic; Triassic**	***Schilderia***
fm: Chinle	**Gray**	**43 cm**

Arizona, USA
Mesozoic; Triassic
unidentified fm: Chinle
Daniels 9 cm tall

Arizona, USA [Winslow]
Mesozoic; Triassic
conifer fm: Chinle
Gray 10.5 cm

Arizona, USA [Winslow]
Mesozoic; Triassic
conifer fm: Chinle
Jones 20.5 cm

Arizona, USA
Mesozoic; Triassic
***Woodworthia* fm: Chinle**
Jones 7.5 cm

Louisiana

Petrified *Palmoxylon* from Louisiana is found in a subtle range of colors from tan to gold, white to black, and pink to violet. The highly silicified specimens ring like a china plate when tapped. Louisiana petrified palm wood has been the state fossil since 1976. These specimens are from the Catahoula Formation, which is dated to the Oligocene or Miocene Epoch of the Tertiary Period. This formation spans the Texas/Louisiana border. The palms of Louisiana grew in a tropical, low-lying, coastal plain.

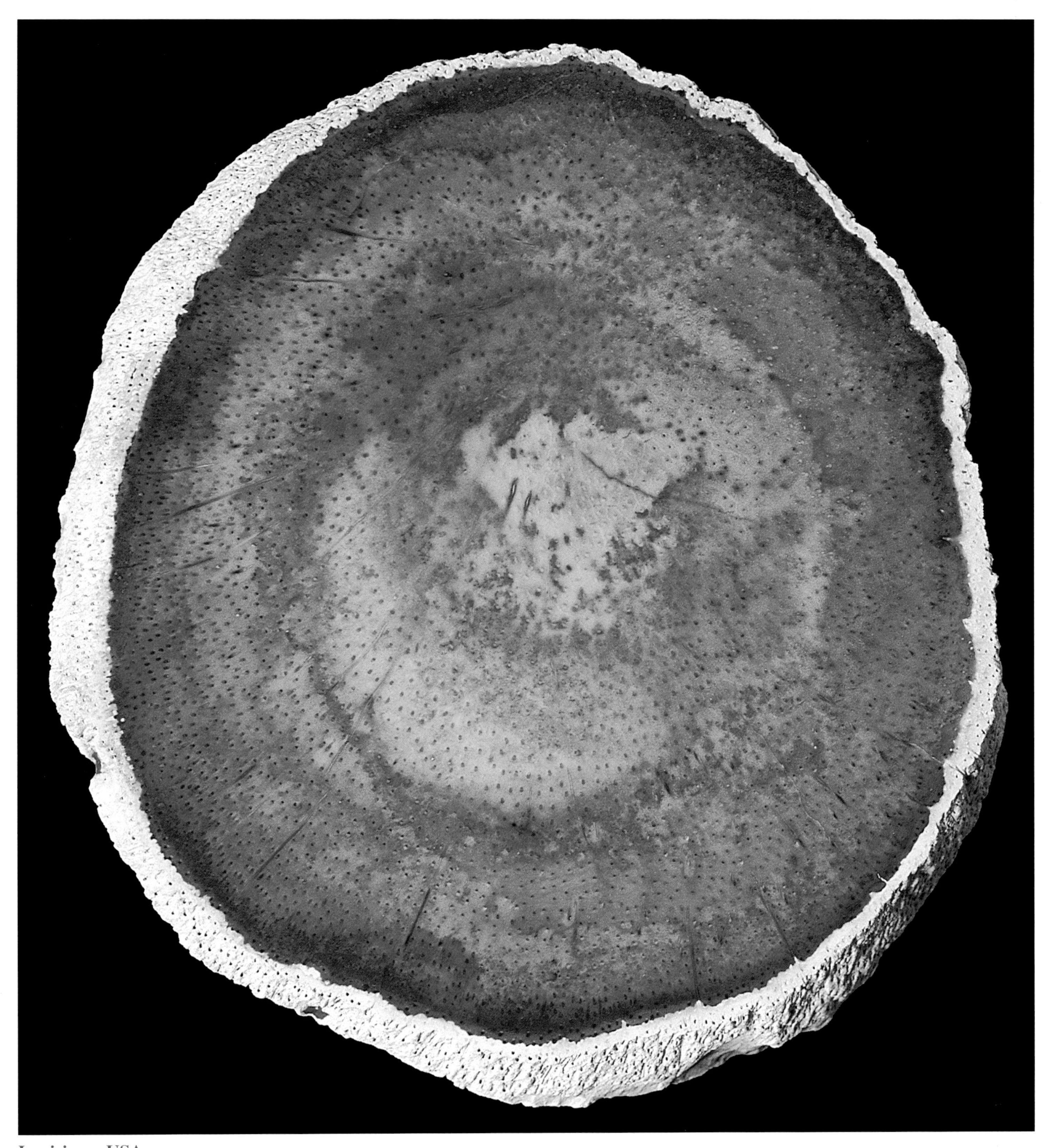

Louisiana, USA
Cenozoic; Tertiary; Oligocene
Palmoxylon **fm: Catahoula**
Akin **19.5 cm**

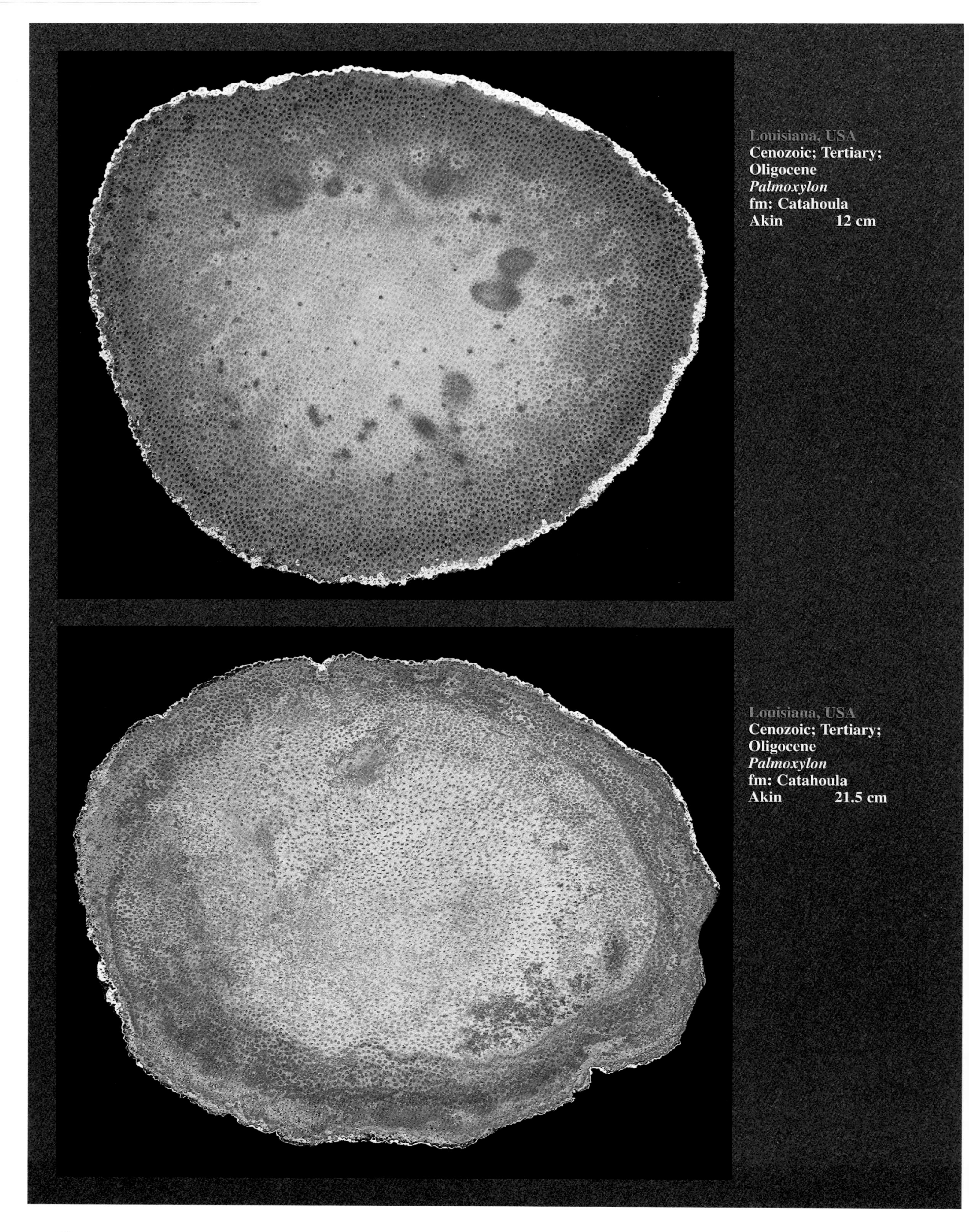

Louisiana, USA
Cenozoic; Tertiary;
Oligocene
Palmoxylon
fm: Catahoula
Akin 12 cm

Louisiana, USA
Cenozoic; Tertiary;
Oligocene
Palmoxylon
fm: Catahoula
Akin 21.5 cm

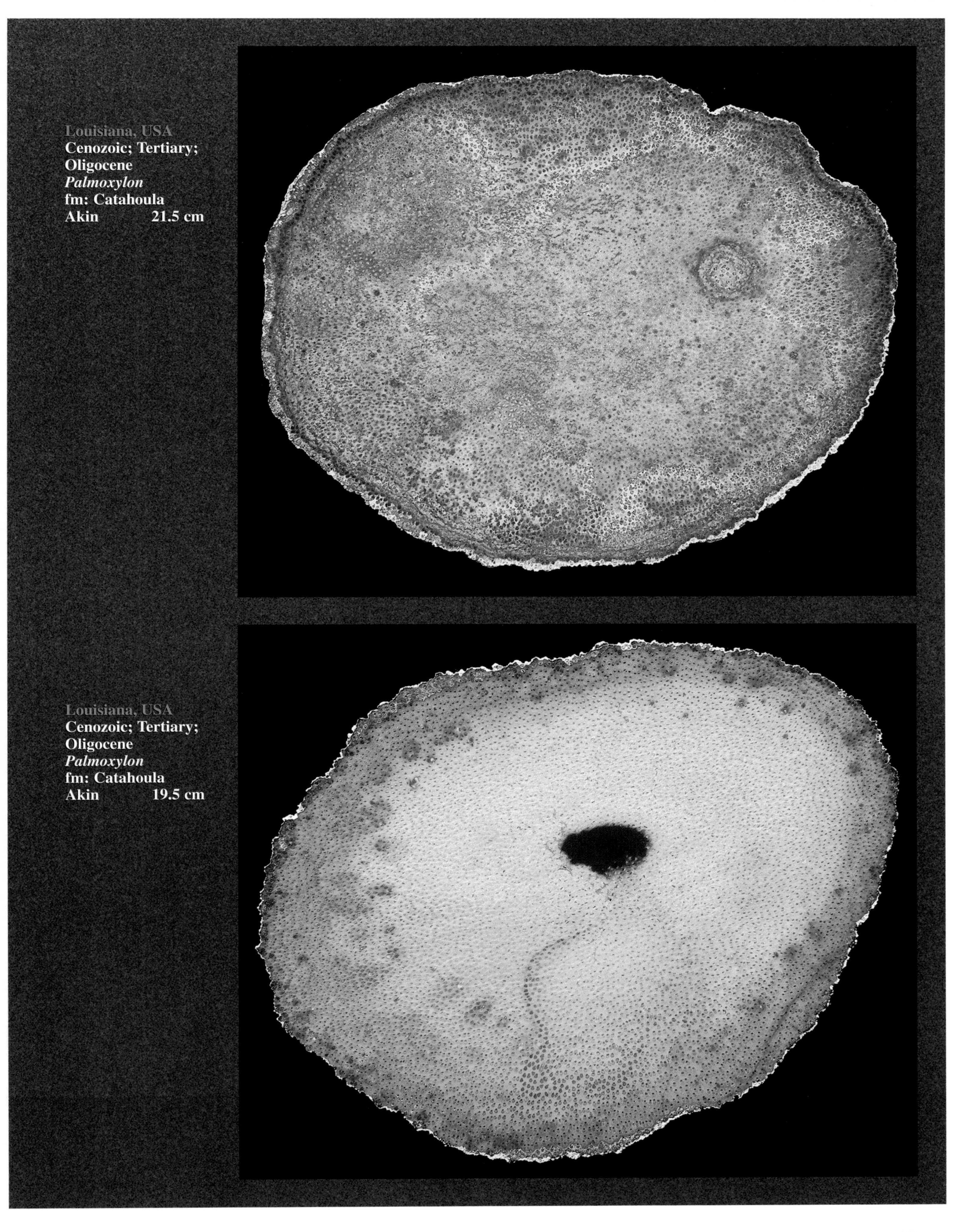

Louisiana, USA
Cenozoic; Tertiary; Oligocene
Palmoxylon
fm: Catahoula
Akin 21.5 cm

Louisiana, USA
Cenozoic; Tertiary; Oligocene
Palmoxylon
fm: Catahoula
Akin 19.5 cm

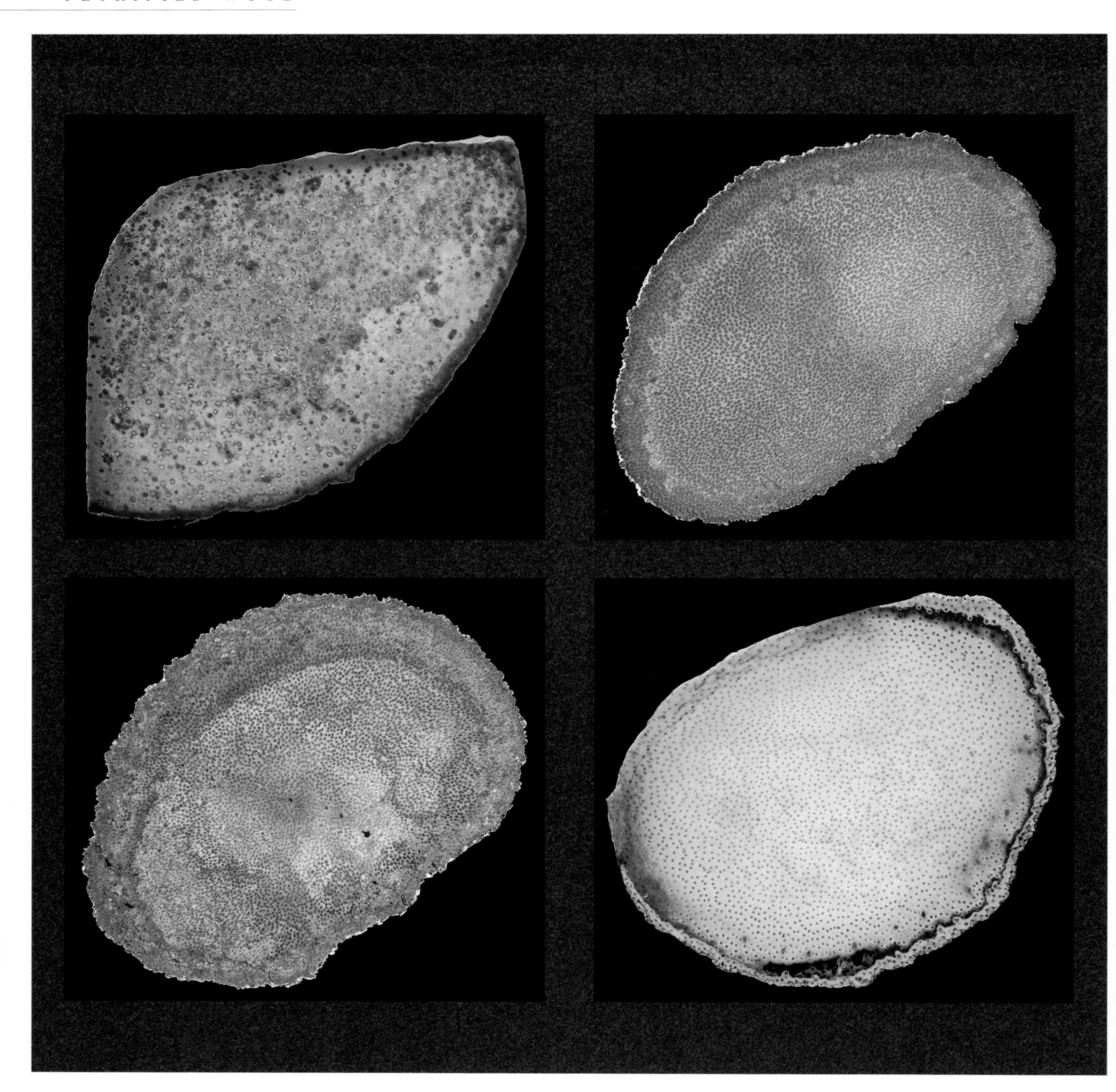

Louisiana, USA
Cenozoic; Tertiary; Oligocene
Palmoxylon
fm: Catahoula
Daniels 14.5 cm

Louisiana, USA
Cenozoic; Tertiary; Oligocene
Palmoxylon
fm: Catahoula
Akin 17 cm

Louisiana, USA
Cenozoic; Tertiary; Oligocene
Palmoxylon
fm: Catahoula
Akin 17 cm

Louisiana, USA
Cenozoic; Tertiary; Oligocene
Palmoxylon
fm: Catahoula
Daniels 22 cm

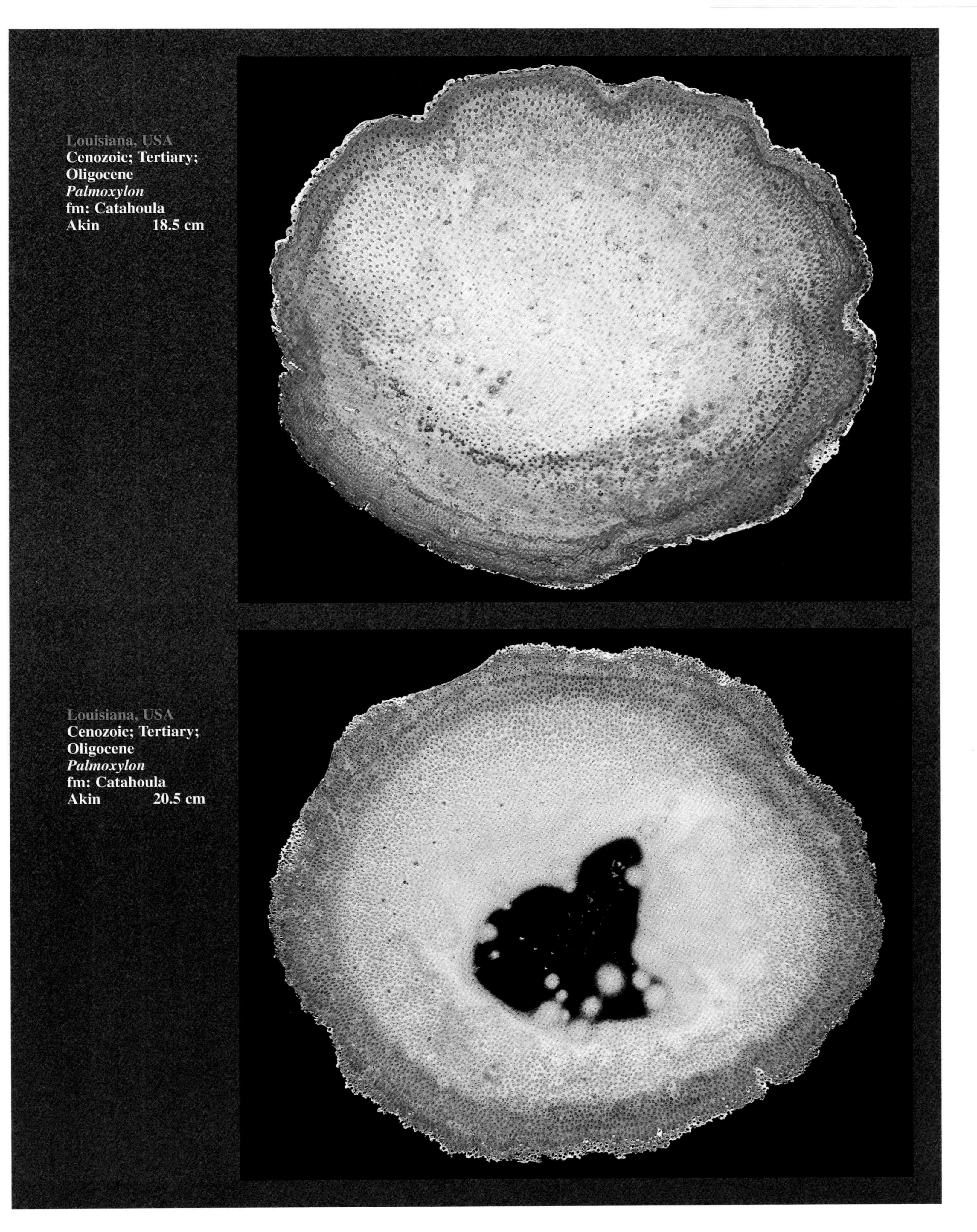

Louisiana, USA
Cenozoic; Tertiary;
Oligocene
Palmoxylon
fm: Catahoula
Akin 18.5 cm

Louisiana, USA
Cenozoic; Tertiary;
Oligocene
Palmoxylon
fm: Catahoula
Akin 20.5 cm

NEVADA

Petrified wood from Nevada comes in a variety of distinctive types. Red petrified conifers, often juniper, from Cherry Creek are easily recognized as are the blue specimens from Hubbard Basin. Opalized wood from Virgin Valley is in a class of its own. The petrified wood in the Oregon section from McDermitt is found on both sides of the Oregon/Nevada border; the town of McDermitt is actually in Nevada.

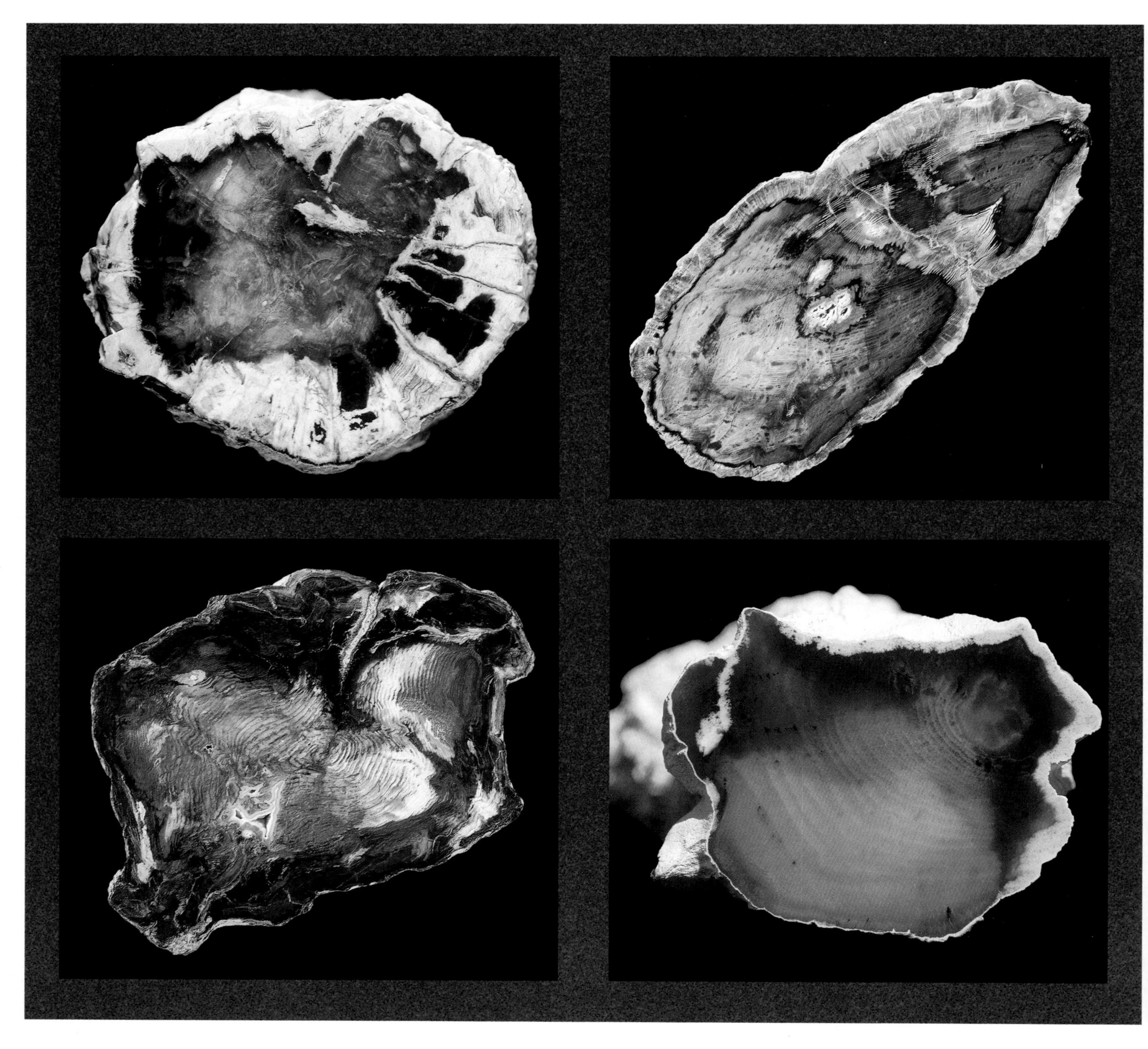

(UPPER LEFT)
Nevada, USA
[Elko County]
Cenozoic; Tertiary; Miocene
conifer
Daniels 11 cm

(LOWER LEFT)
Nevada, USA
[Cherry Creek]
Cenozoic; Tertiary; Miocene
conifer
Rigel 20 cm

(UPPER RIGHT)
Nevada, USA
[Tuscarora]
Cenozoic; Tertiary; Miocene
conifer
Daniels 19.5 cm

(LOWER RIGHT)
Nevada, USA
[Little Humboldt River]
Cenozoic; Tertiary
***Salix* [willow]**
Daniels 5 cm

(TOP)
Nevada, USA [Elko County]
Cenozoic; Tertiary; Miocene
***Juniperus* [juniper]**
Daniels 30 cm

(LOWER LEFT)
Nevada, USA [Virgin Valley]
Cenozoic; Tertiary; Miocene
unidentified fm: Virgin Valley
Dayvault 4.5 cm

(LOWER RIGHT)
Nevada, USA [Virgin Valley]
Cenozoic; Tertiary; Miocene
unidentified fm: Virgin Valley
Macom 2.5 cm

Nevada, USA
[Cherry Creek]
Cenozoic; Tertiary; Miocene
conifer
Jones 15 cm

Nevada, USA
[Hubbard Basin]
Cenozoic; Tertiary
unidentified
Cataldo 18.5 cm

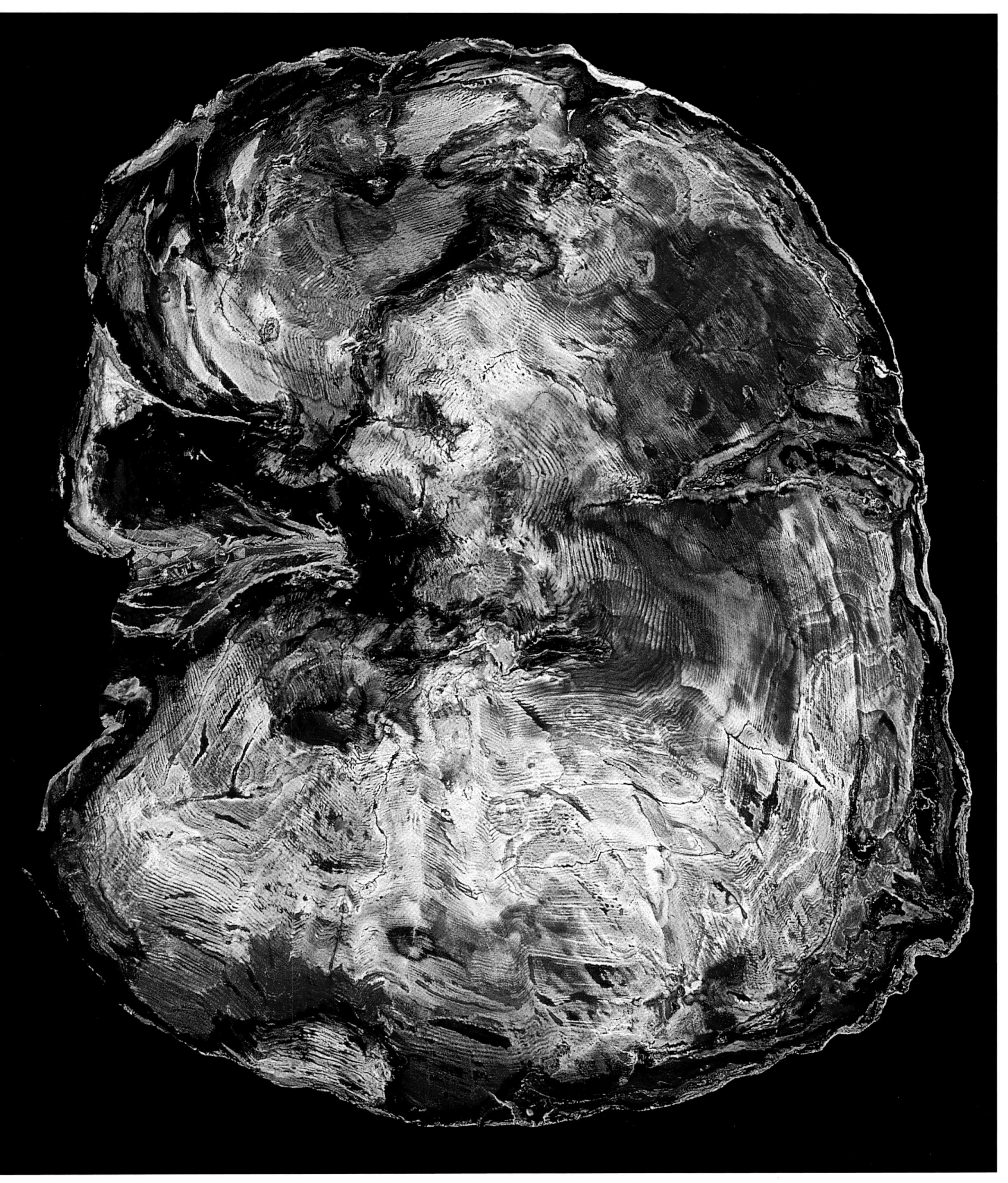

Nevada, USA [Cherry Creek]
Cenozoic; Tertiary; Miocene
conifer
Cataldo 37.5 cm

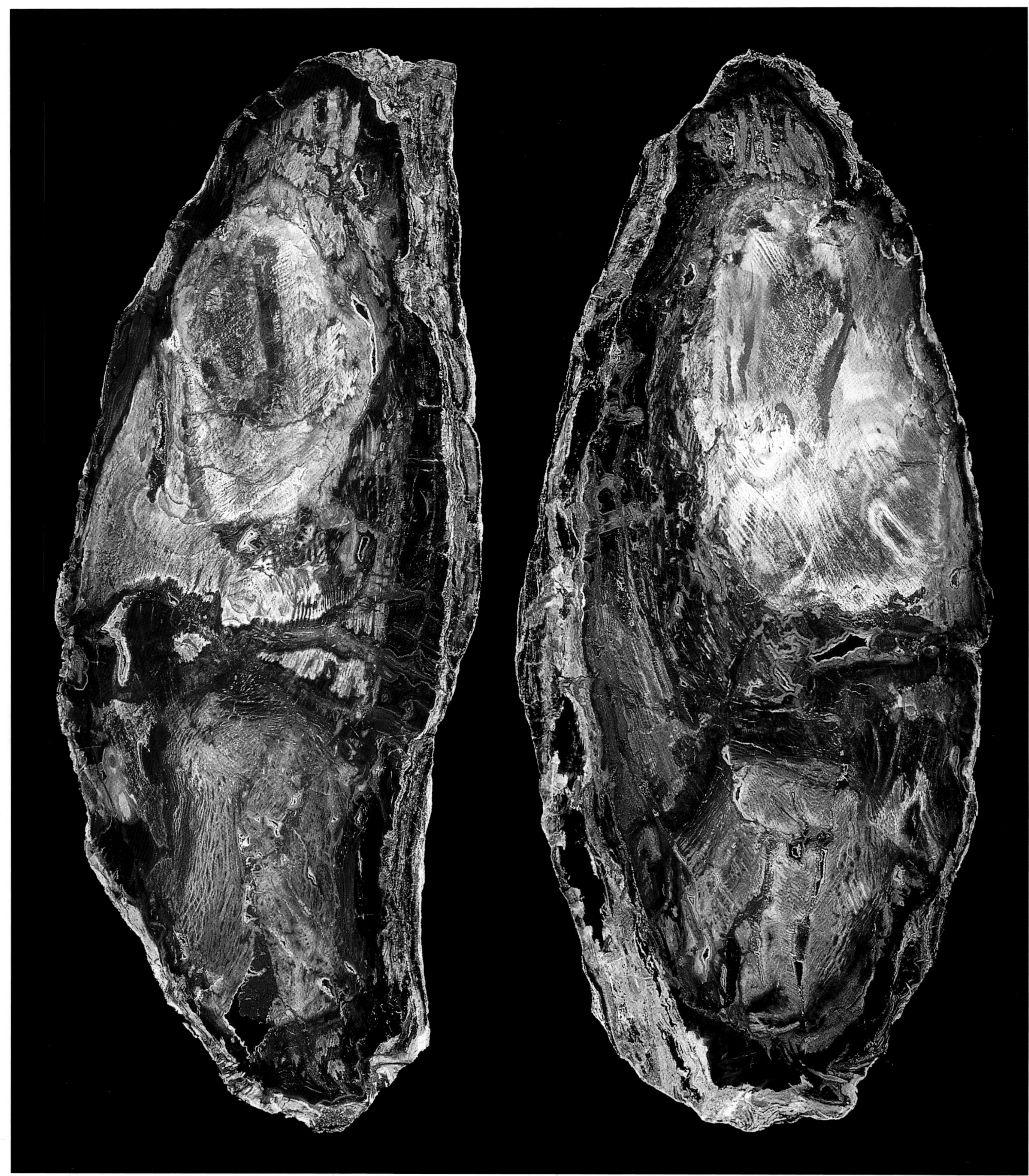

Nevada, USA [Cherry Creek]
Cenozoic; Tertiary; Miocene
***Juniperus* [juniper]**
Daniels 29 cm

Nevada, USA [Cherry Creek]
Cenozoic; Tertiary; Miocene
***Juniperus* [juniper]**
Daniels 29.5 cm

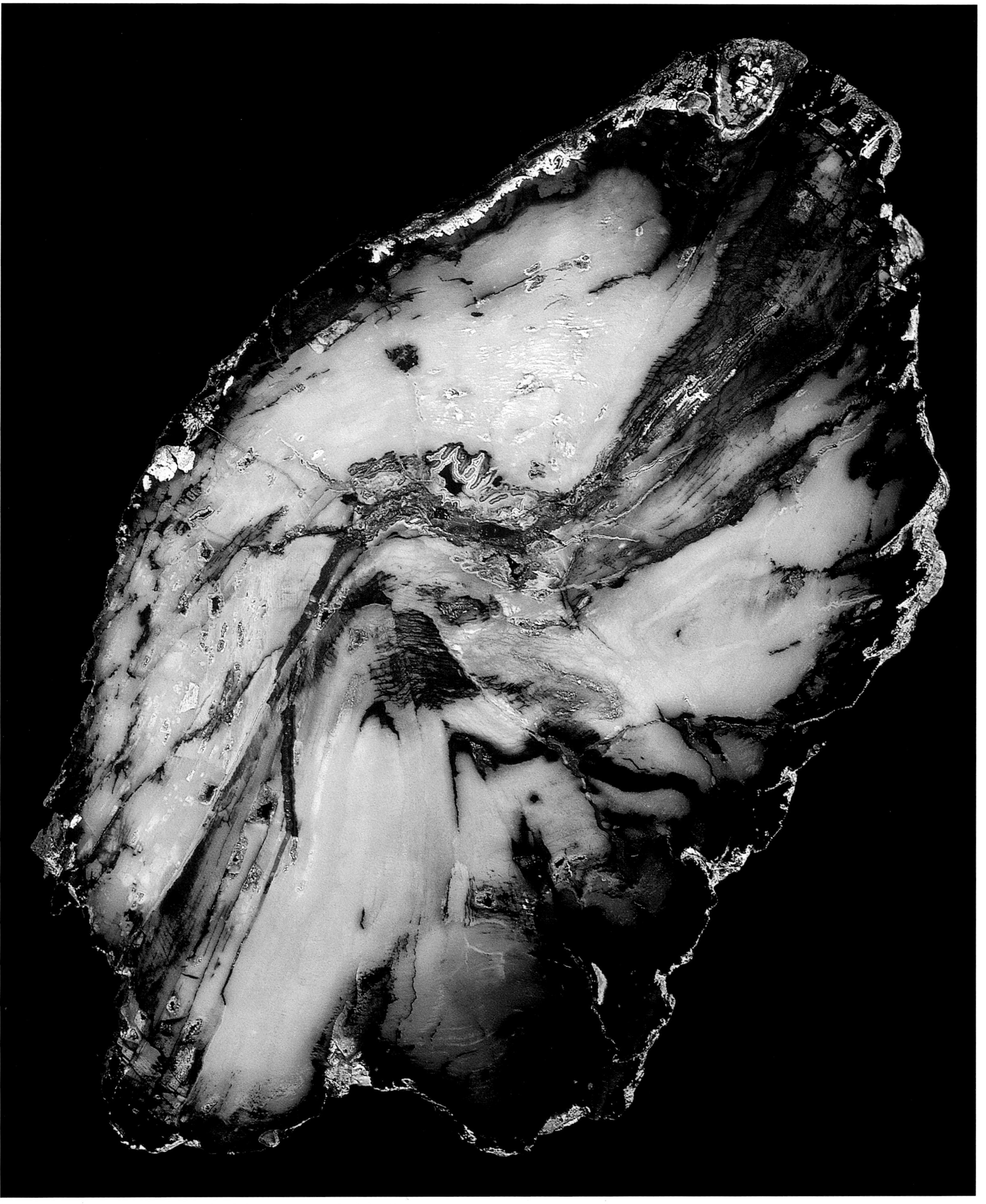

Nevada, USA [Hubbard Basin] — Cenozoic; Tertiary
unidentified — Keller — 32 cm

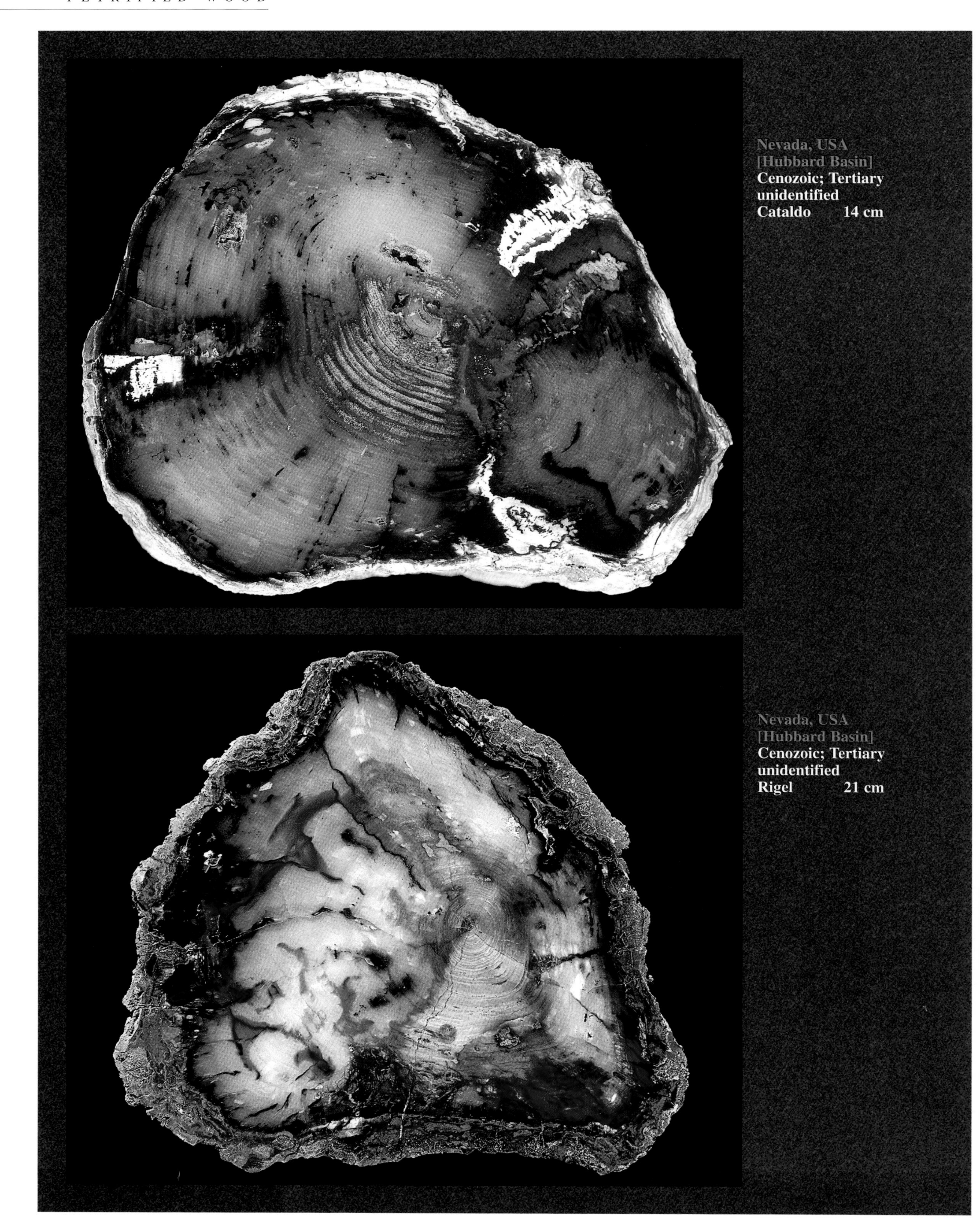

Nevada, USA
[Hubbard Basin]
Cenozoic; Tertiary
unidentified
Cataldo 14 cm

Nevada, USA
[Hubbard Basin]
Cenozoic; Tertiary
unidentified
Rigel 21 cm

OREGON

Woods from Oregon present a vast array of color and variety, thanks primarily to the numerous gymnosperms and angiosperms that prospered during the Tertiary Period and the active volcanism in the region. Petrified wood is found in many places in Oregon, each showing distinctive mineralization. Identification of the genus is often possible by comparison of fossil thin-sections with living woods. Deschutes specimens, from north central Oregon, exhibit remarkable detail. Stinkingwater specimens, from eastern Oregon, are frequently oak permineralized black and brown; small branches are rare, as flaming lava or hot ash evidently consumed smaller trees and branches. Sweet Home, in west central Oregon, is noted for its variety of petrified wood genera.

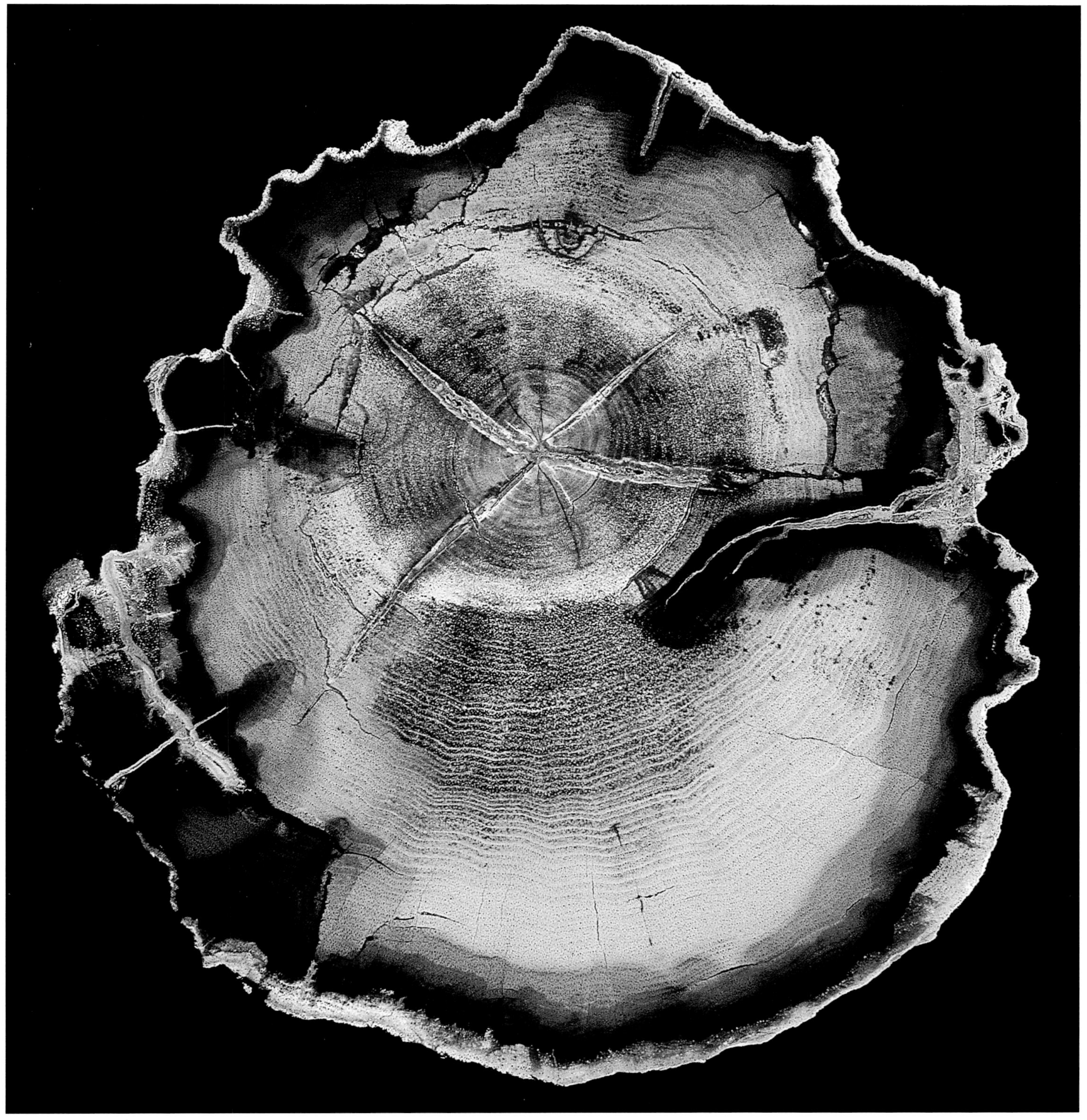

Oregon, USA [Deschutes]
Cenozoic; Tertiary; Miocene/Pliocene
***Carya* [hickory] Daniels 17 cm**

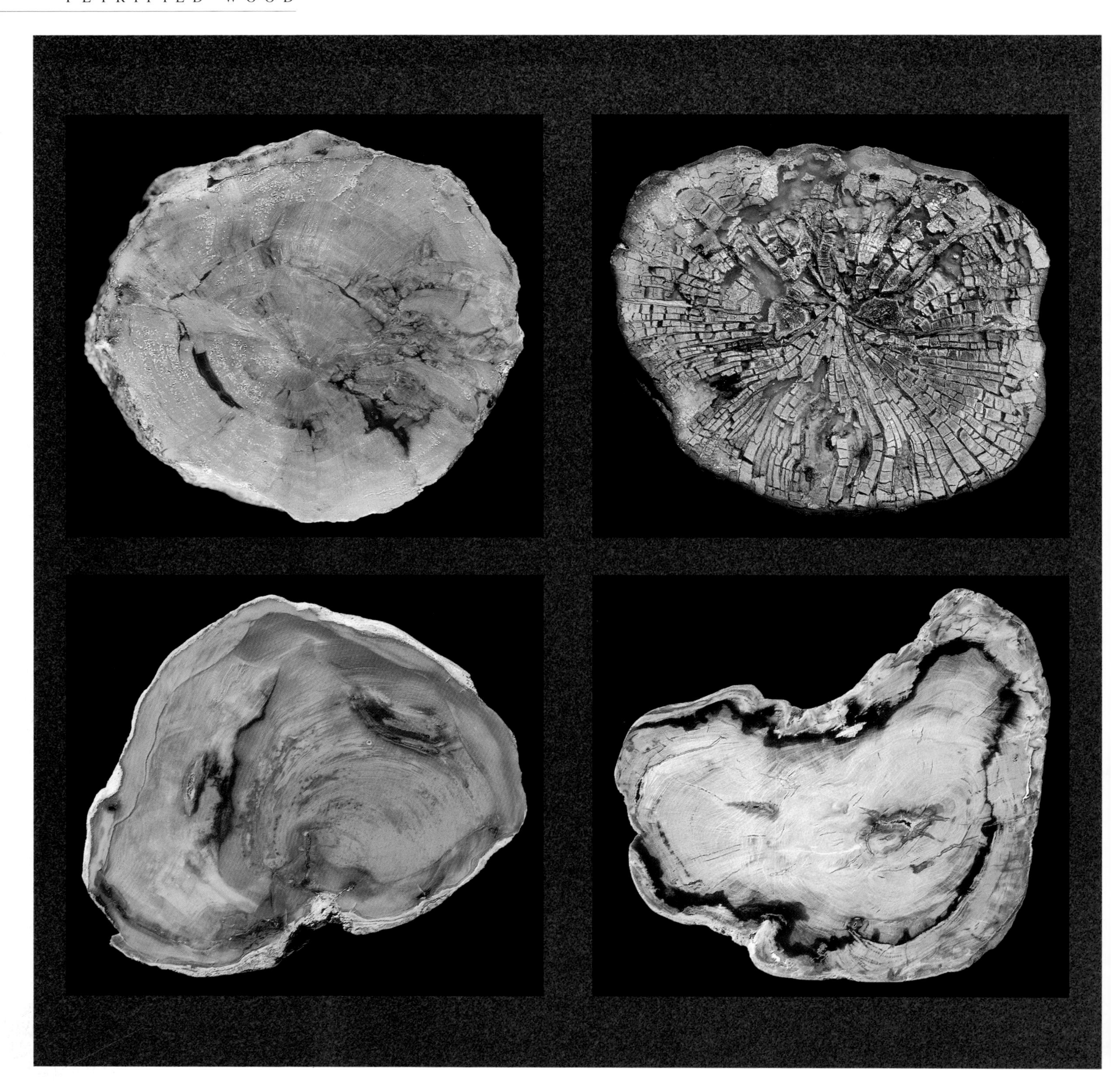

Oregon, USA
Cenozoic; Tertiary
unidentified
Jones 6 cm

Oregon, USA [Rogue River]
Cenozoic; Tertiary
unidentified
Rigel 10 cm

Oregon, USA
Cenozoic; Tertiary
unidentified
Daniels 13 cm

Oregon , USA
Cenozoic; Tertiary
cedar
Daniels 16 cm

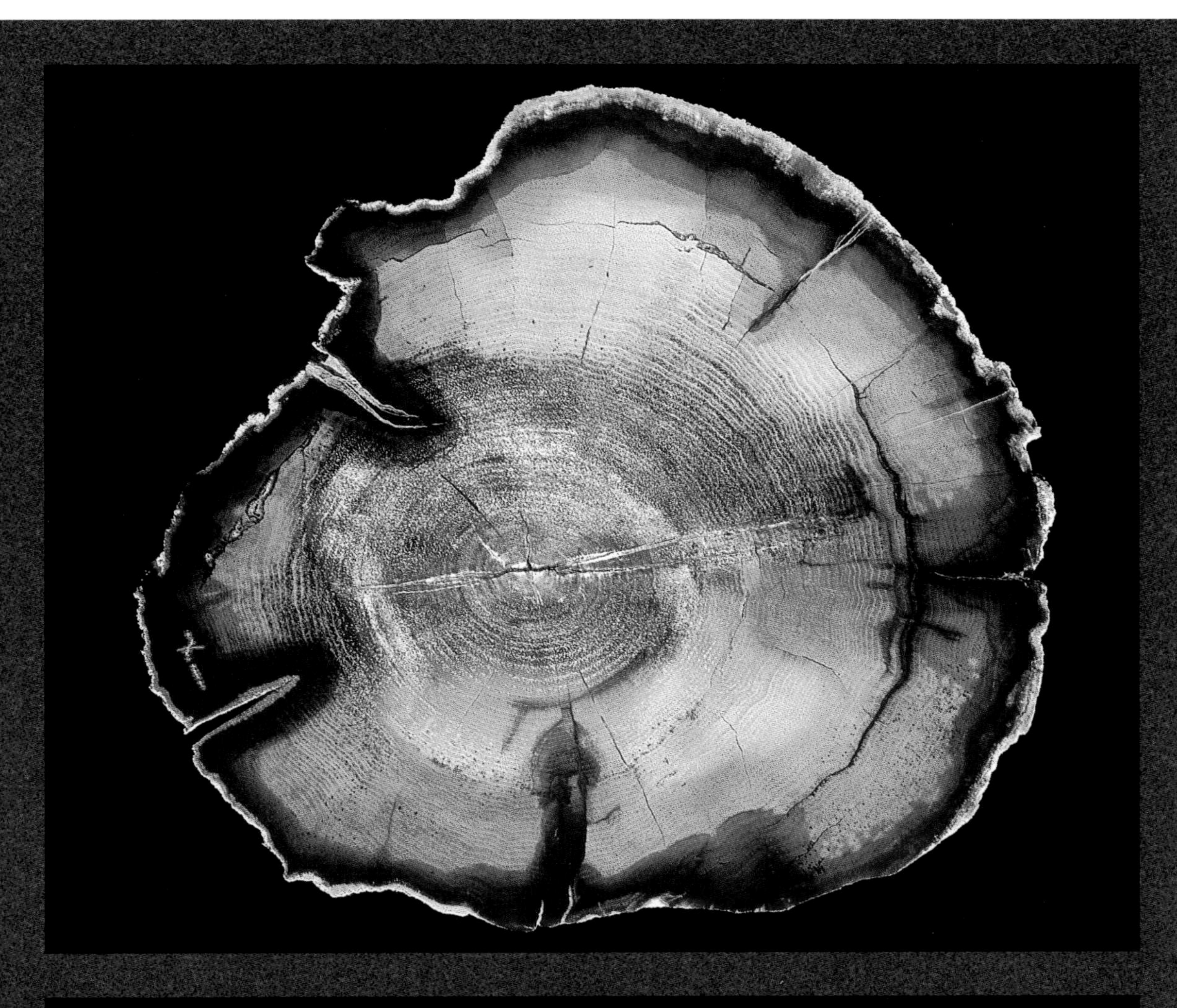

Oregon, USA
[Deschutes]
Cenozoic; Tertiary;
Miocene/Pliocene
***Carya* [hickory]**
Daniels 16 cm

Oregon, USA
[Hampton Butte]
Cenozoic; Tertiary;
Eocene/Oligocene
unidentified
fm: Clarno
Cataldo 12 cm

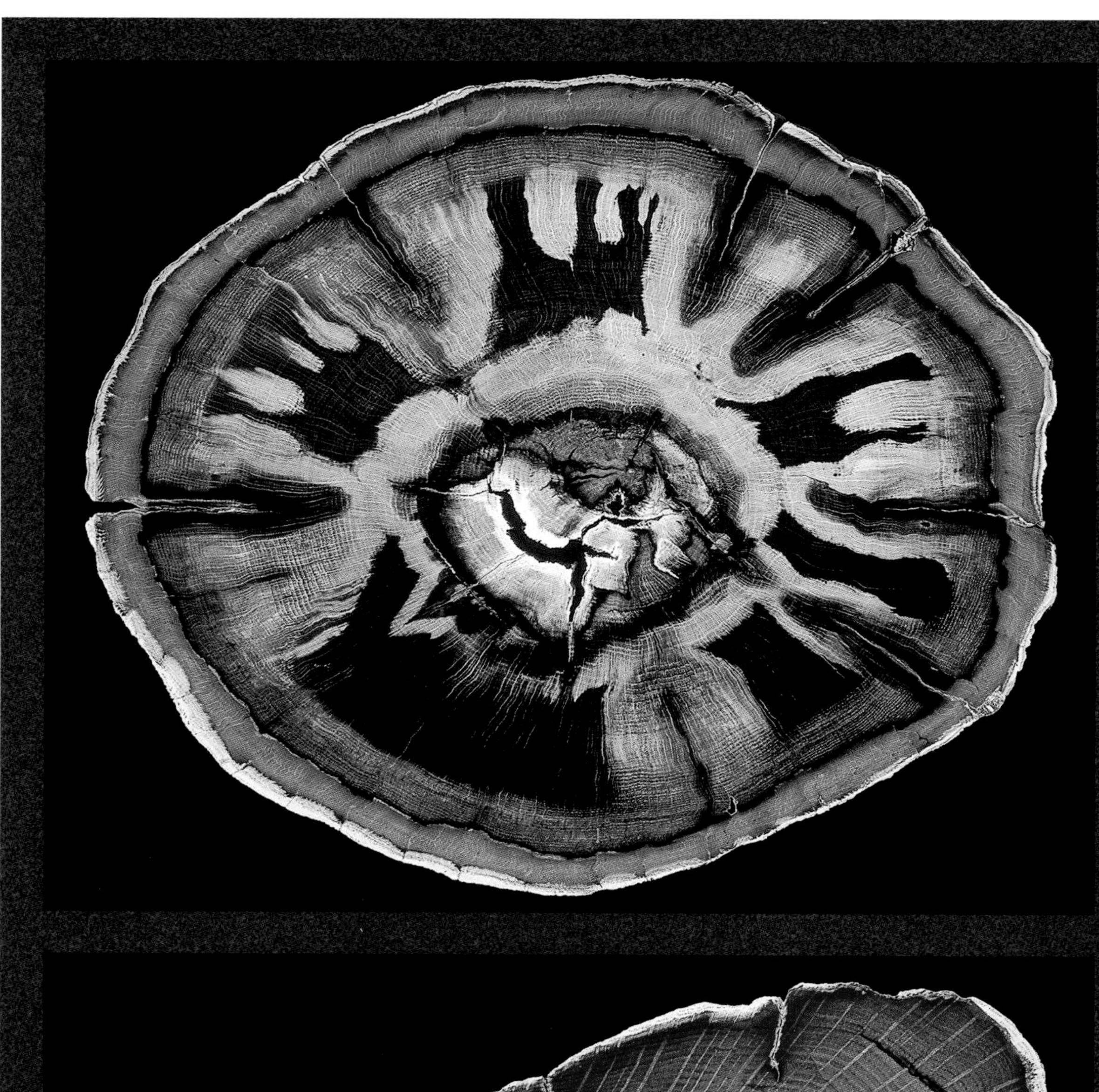

Oregon, USA
[Stinkingwater]
Cenozoic; Tertiary; Miocene
***Quercus* [oak]**
Robertson 41 cm

Oregon, USA
[Stinkingwater]
Cenozoic; Tertiary; Miocene
***Quercus* [oak]**
Cataldo 17.5 cm

Oregon, USA [Deschutes]
Cenozoic; Tertiary; Miocene/Pliocene
***Quercus* [oak; double-hearted]**
Rigel 30.5 cm

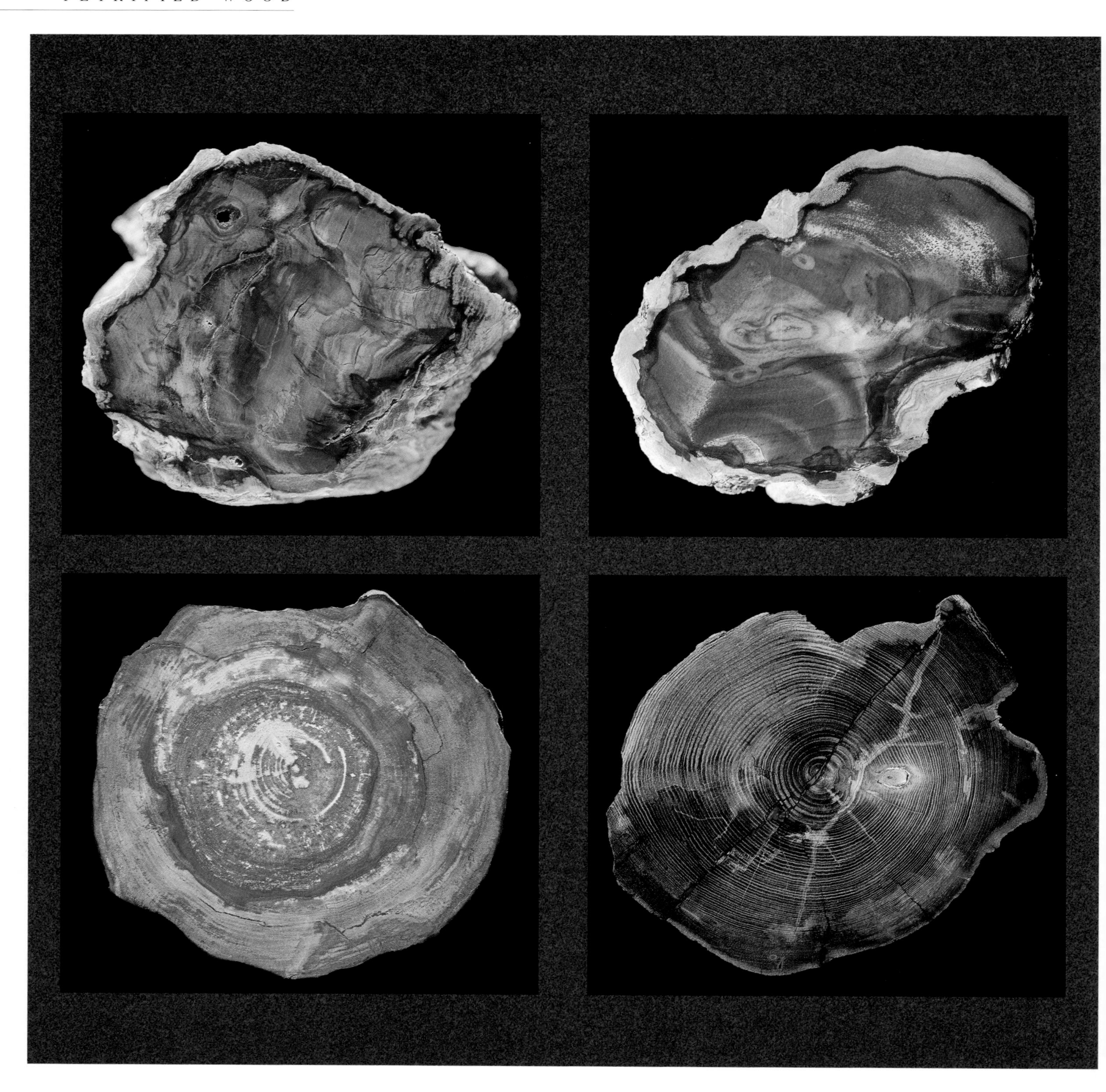

Oregon, USA
Cenozoic; Tertiary
unidentified
Daniels 7.5 cm

Oregon, USA
Cenozoic; Tertiary
unidentified
Daniels 8.5 cm

Oregon, USA [Sweet Home]
Cenozoic; Tertiary; Oligocene/Miocene
unidentified
Daniels 13 cm

Oregon, USA [Sweet Home]
Cenozoic; Tertiary; Oligocene/Miocene
unidentified
Daniels 11 cm

Oregon, USA
[Sweet Home]
Cenozoic; Tertiary;
Oligocene/Miocene
unidentified
Daniels 23 cm

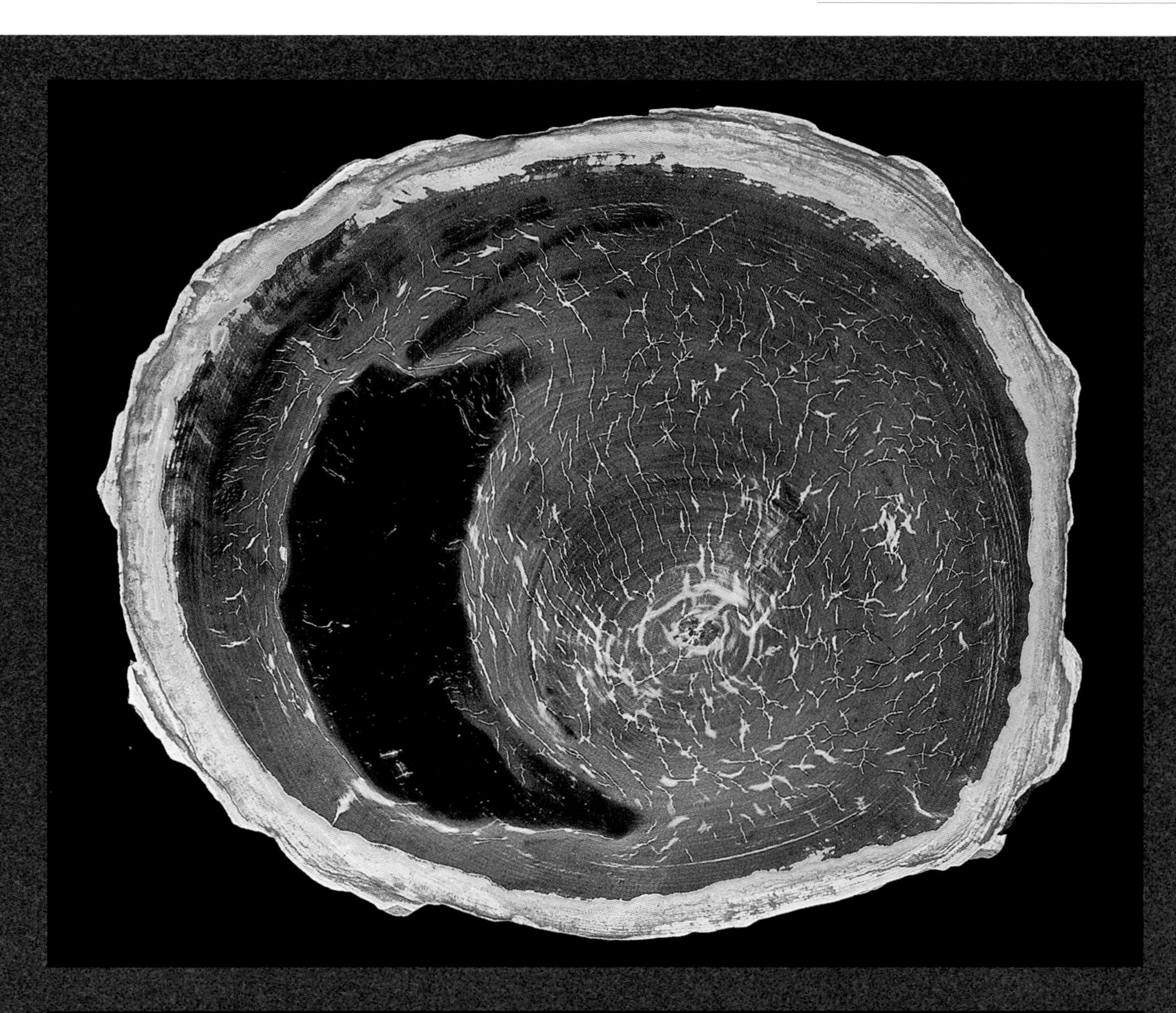

Oregon, USA
[Swartz Canyon]
Cenozoic; Tertiary
***Quercus* [oak]**
Rigel 15 cm

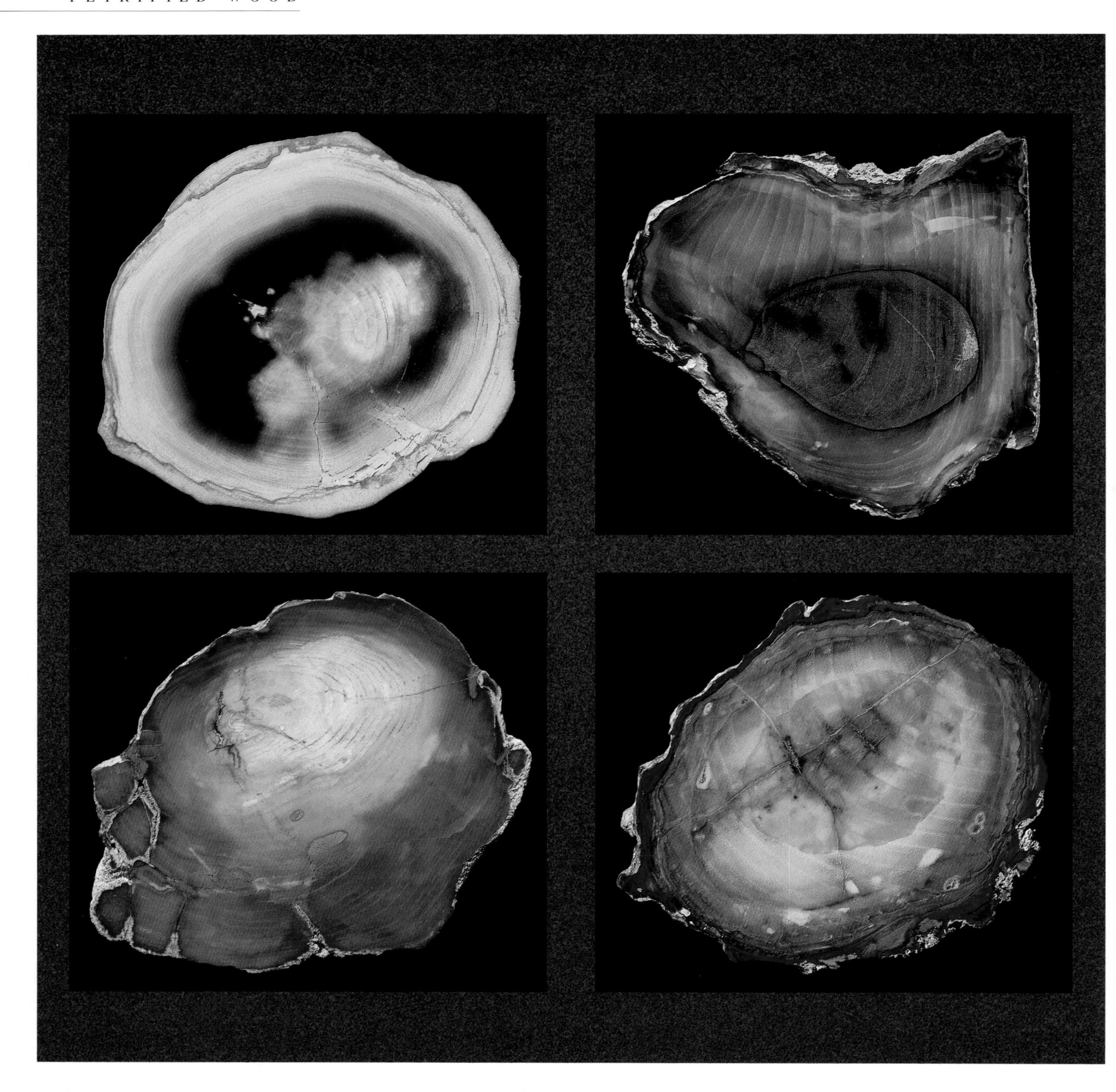

Oregon, USA [Sweet Home]
Cenozoic; Tertiary; Oligocene/Miocene
unidentified
Daniels 8.5 cm

Oregon, USA [Grassy Mountain]
Cenozoic; Tertiary; Miocene
unidentified
Daniels 15 cm

Oregon, USA [Grassy Mountain]
Cenozoic; Tertiary; Miocene
unidentified
Cataldo 13.5 cm

Oregon, USA [Grassy Mountain]
Cenozoic; Tertiary; Miocene
unidentified
Rigel 15 cm

Oregon, USA [Deschutes]
Cenozoic; Tertiary; Miocene/Pliocene
***Quercus* [oak] Robertson 27.5 cm**

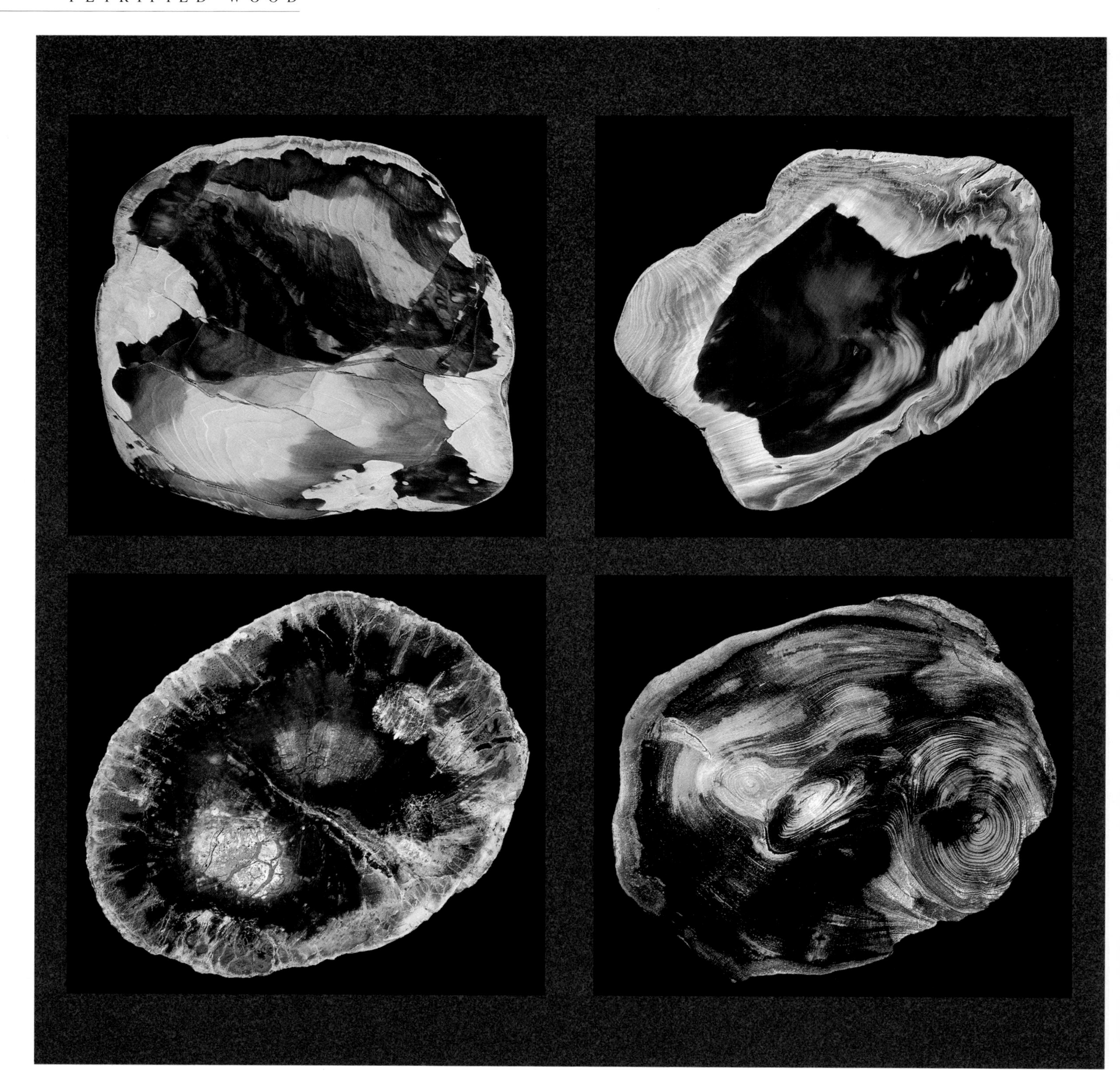

Oregon, USA [Rogers Mountain]
Cenozoic; Tertiary
unidentified
Daniels 16.5 cm

Oregon, USA [Rogers Mountain]
Cenozoic; Tertiary
unidentified
Daniels 16 cm

Oregon, USA [Swartz Canyon]
Cenozoic; Tertiary
unidentified
Daniels 10 cm

Oregon, USA [Knutson Creek]
Cenozoic; Tertiary
***Quercus* [oak; triple-hearted]**
Rigel 10 cm

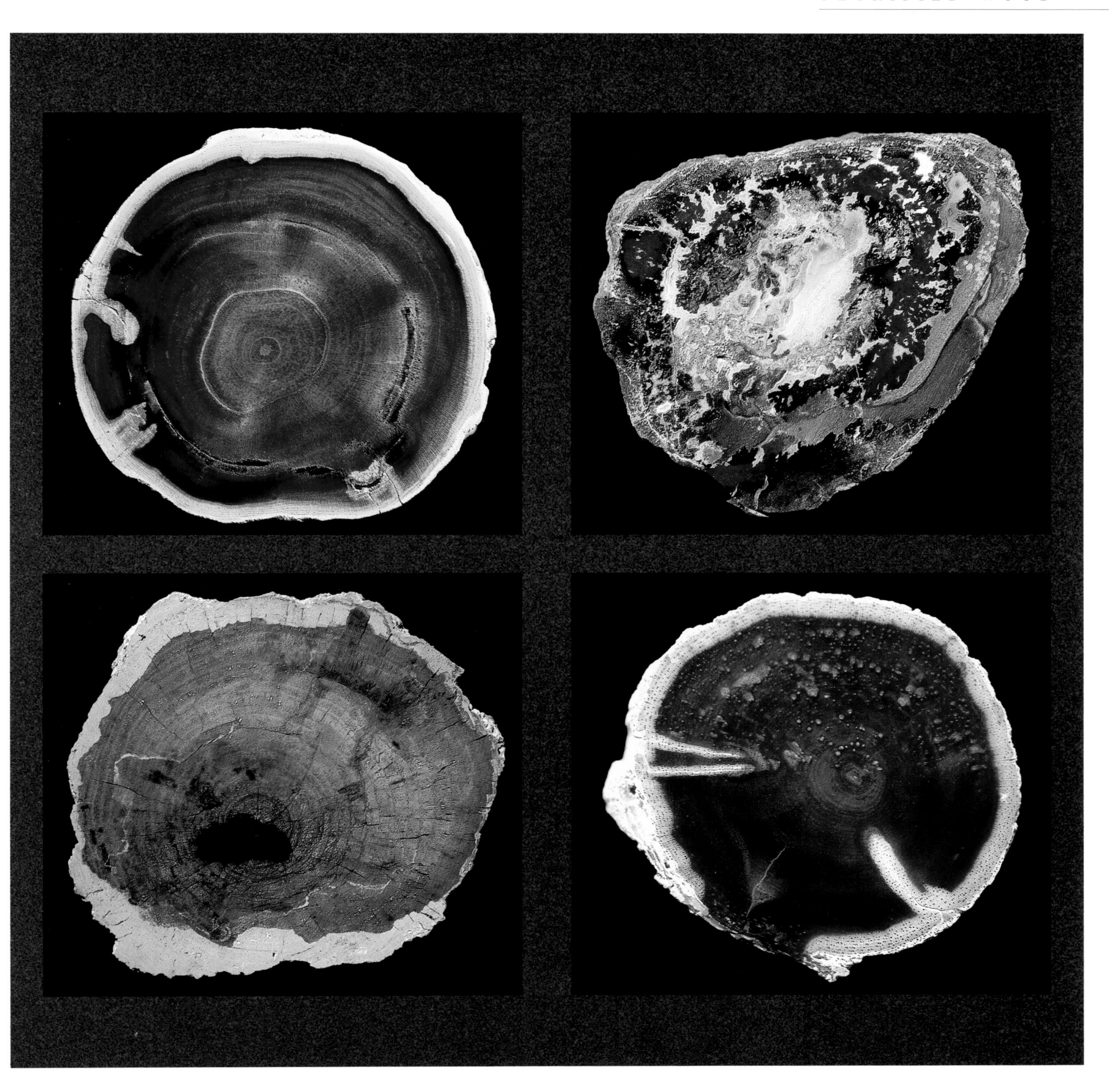

Oregon, USA [McDermitt]
Cenozoic; Tertiary; Miocene
unidentified fm: Trout Creek
Jones 17.5 cm

Oregon, USA [Trail Creek]
Cenozoic; Tertiary
unidentified
Rigel 7.5 cm

Oregon, USA [McDermitt]
Cenozoic; Tertiary; Miocene
***Robinia* [locust]**
Daniels 9 cm

Oregon, USA
Cenozoic; Tertiary
unidentified
Daniels 5.5 cm

UTAH

Variations within Jurassic petrified woods from Utah foster an appreciation for the incredible mechanisms that led to petrification. Branches in Yellow Cat are gems of carnelian agate, while those from the Henry Mountains, just seventy-five miles southwest, are often permineralized yellow and black. Good quality specimens from Utah are found primarily in the Triassic age Chinle and Jurassic age Morrison formations.

The area noted in the captions as Henry Mountains comprises a large expanse extending from north of Hanksville south to the Arizona border. South of Boulder and west of Capitol Reef National Park is the Circle Cliffs area, which was included in Grand Staircase–Escalante National Monument. The Yellow Cat area, an old uranium mining district, is located within the triangle formed by Crescent Junction, Cisco, and Moab.

(UPPER LEFT)
Utah, USA
[Henry Mountains]
Mesozoic; Jurassic
conifer fm: Morrison
Jones 9 cm

(LOWER LEFT)
Utah, USA
[Henry Mountains]
Mesozoic; Triassic
conifer fm: Chinle
Daniels 6 cm

(UPPER RIGHT)
Utah, USA
[Henry Mountains]
Mesozoic; Jurassic
unidentified fm: Morrison
Jones 8.5 cm

(LOWER RIGHT)
Utah, USA
[Henry Mountains]
Mesozoic; Jurassic
unidentified fm: Morrison
Jones 7 cm

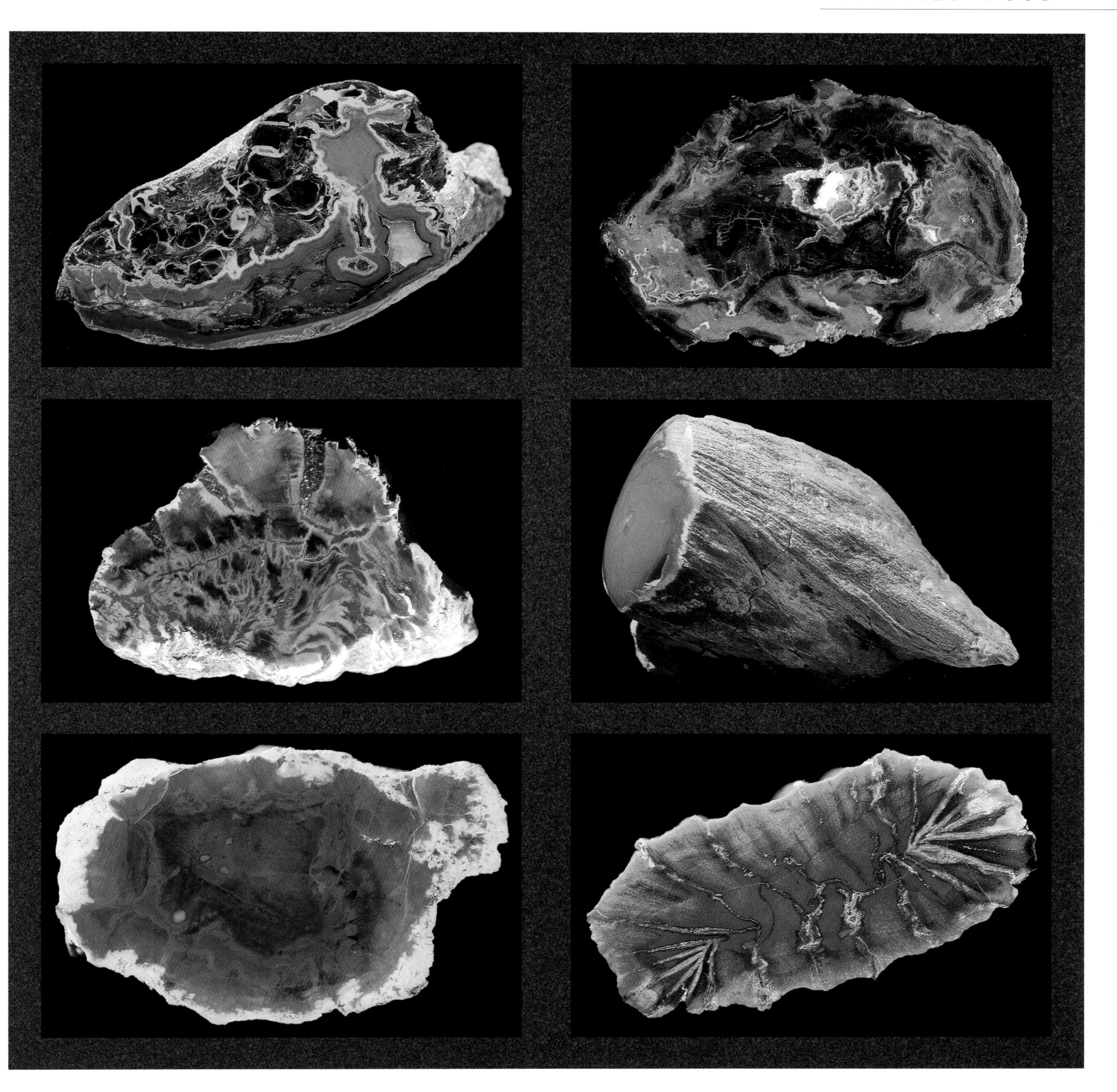

Utah, USA [Henry Mountains]
Mesozoic; Jurassic
unidentified fm: Morrison
Jones 8 cm

Utah, USA [Henry Mountains]
Mesozoic; Triassic
conifer fm: Chinle
Daniels 8.5 cm

Utah, USA [Escalante]
Mesozoic; Jurassic
conifer fm: Morrison
Daniels 6 cm

Utah, USA [Four Corners]
Mesozoic; Jurassic
conifer fm: Morrison
Jones 10.5 cm

Utah, USA [Henry Mountains]
Mesozoic; Jurassic
unidentified fm: Morrison
Jones 6 cm long

Utah, USA [Circle Cliffs]
Mesozoic; Triassic
unidentified fm: Chinle
Daniels 4.5 cm

Utah, USA
[Henry Mountains]
Mesozoic; Jurassic
unidentified
fm: Morrison
Jones 10 cm

Utah, USA
[Henry Mountains]
Mesozoic; Jurassic
conifer
fm: Morrison
Jones 10.5 cm

Utah, USA
[Henry Mountains]
Mesozoic; Jurassic
conifer
fm: Morrison
Jones 7 cm

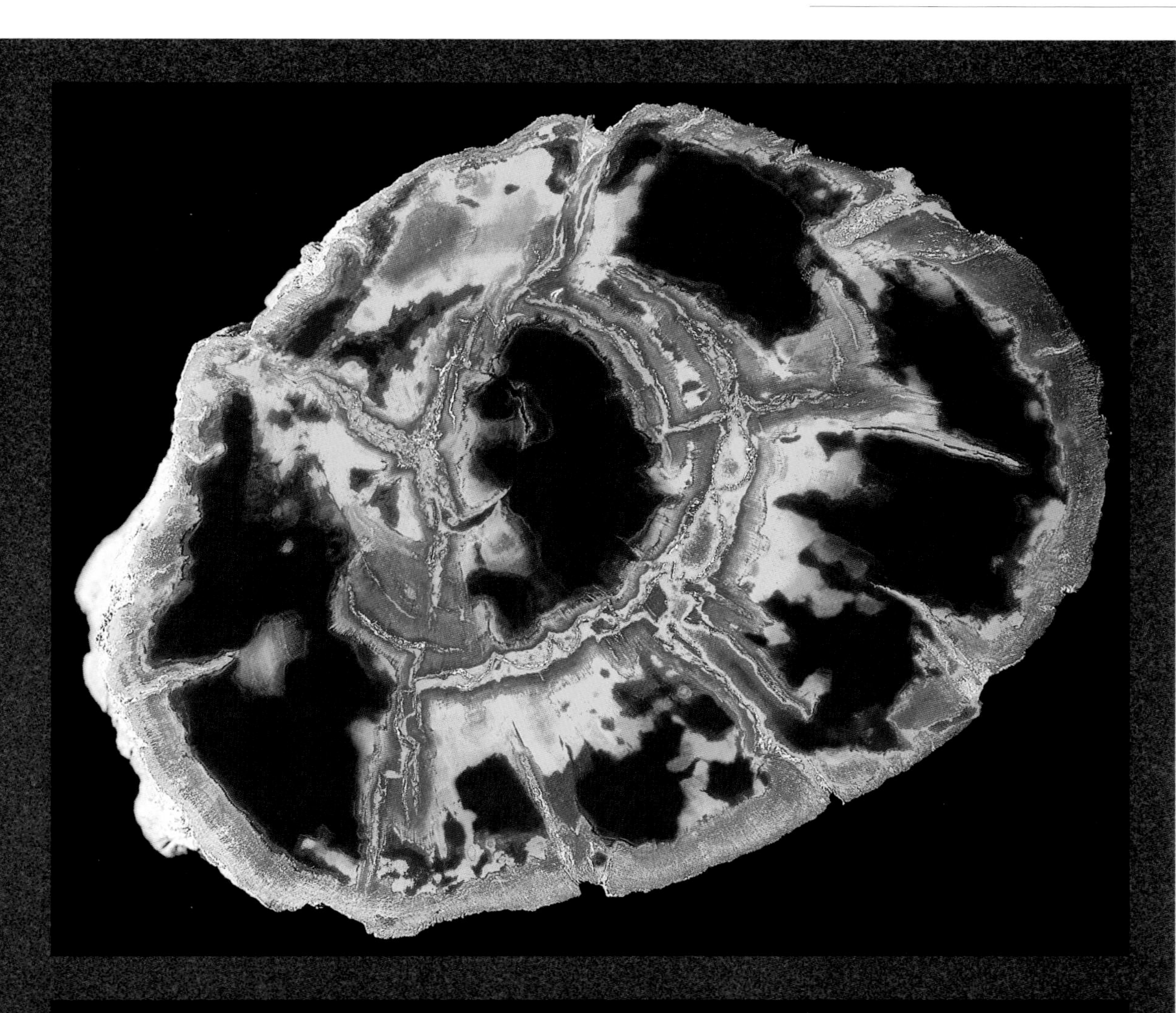

Utah, USA
[Henry Mountains]
Mesozoic; Jurassic
unidentified
fm: Morrison
Jones 15 cm

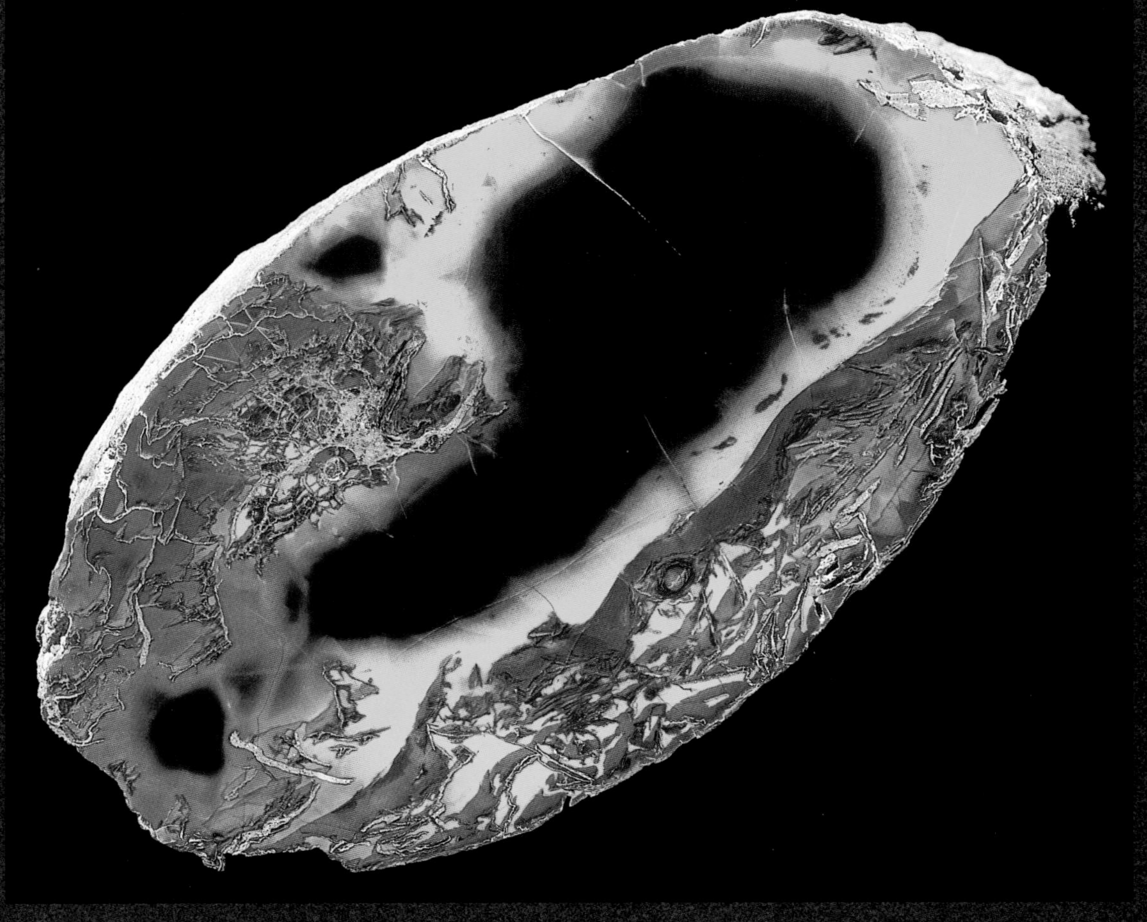

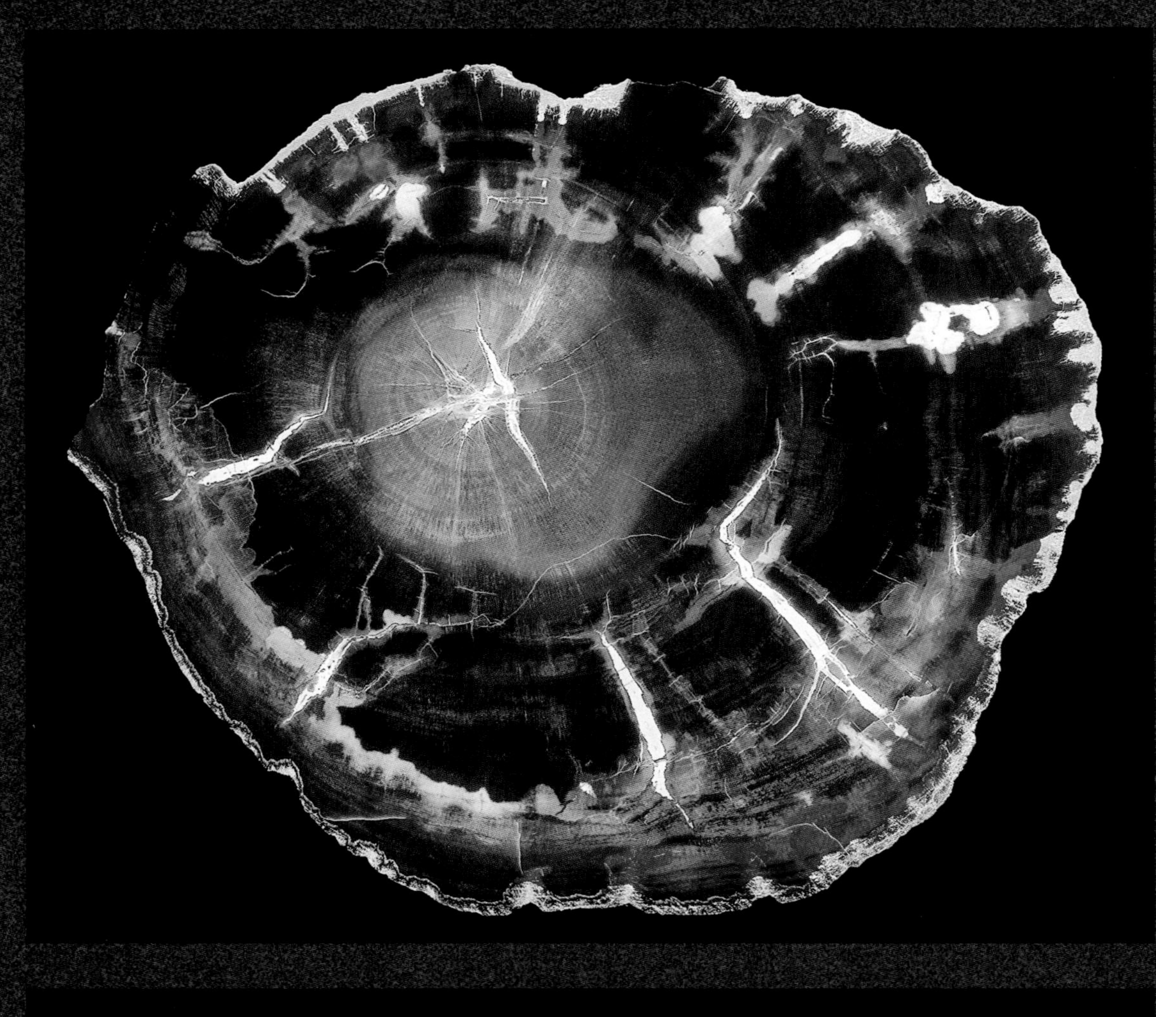

Utah, USA [Paria River]
Mesozoic; Jurassic
conifer
fm: Morrison
Daniels 18 cm

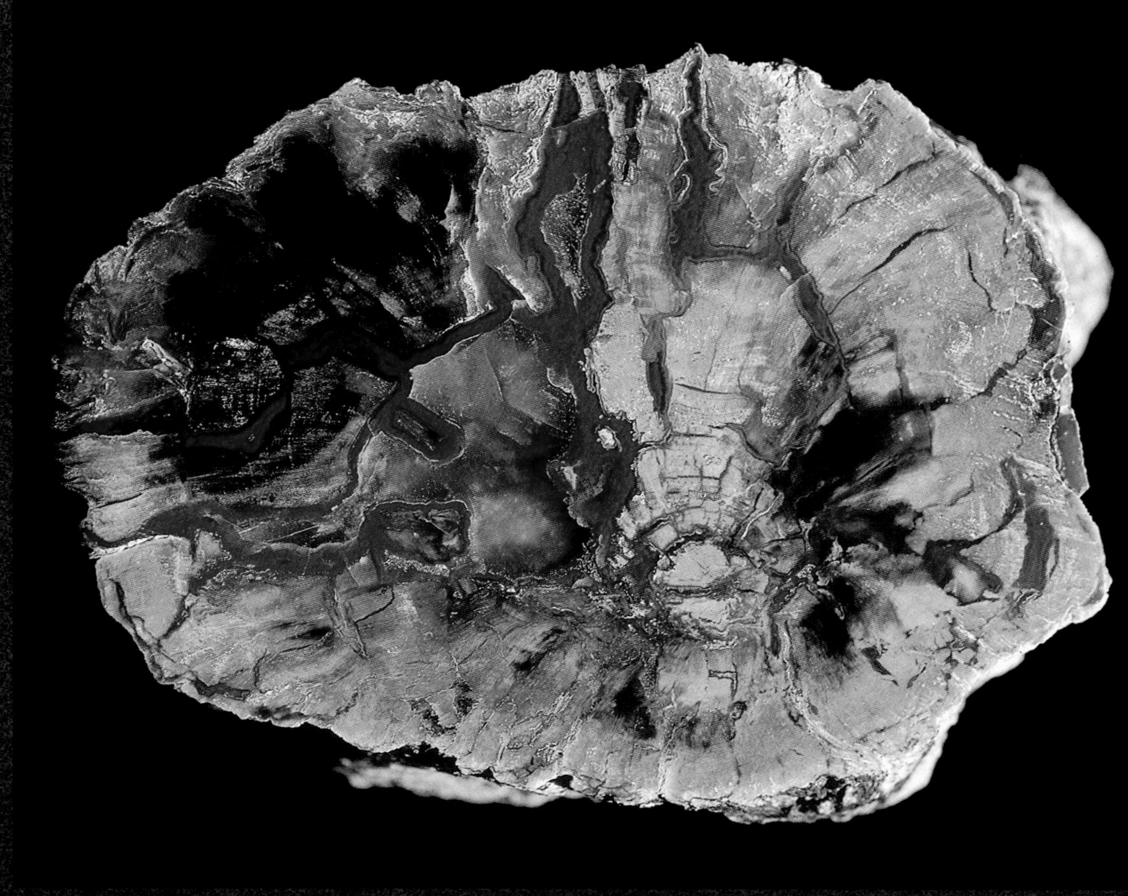

Utah, USA
[Henry Mountains]
Mesozoic; Jurassic
conifer
fm: Morrison
Jones 8 cm

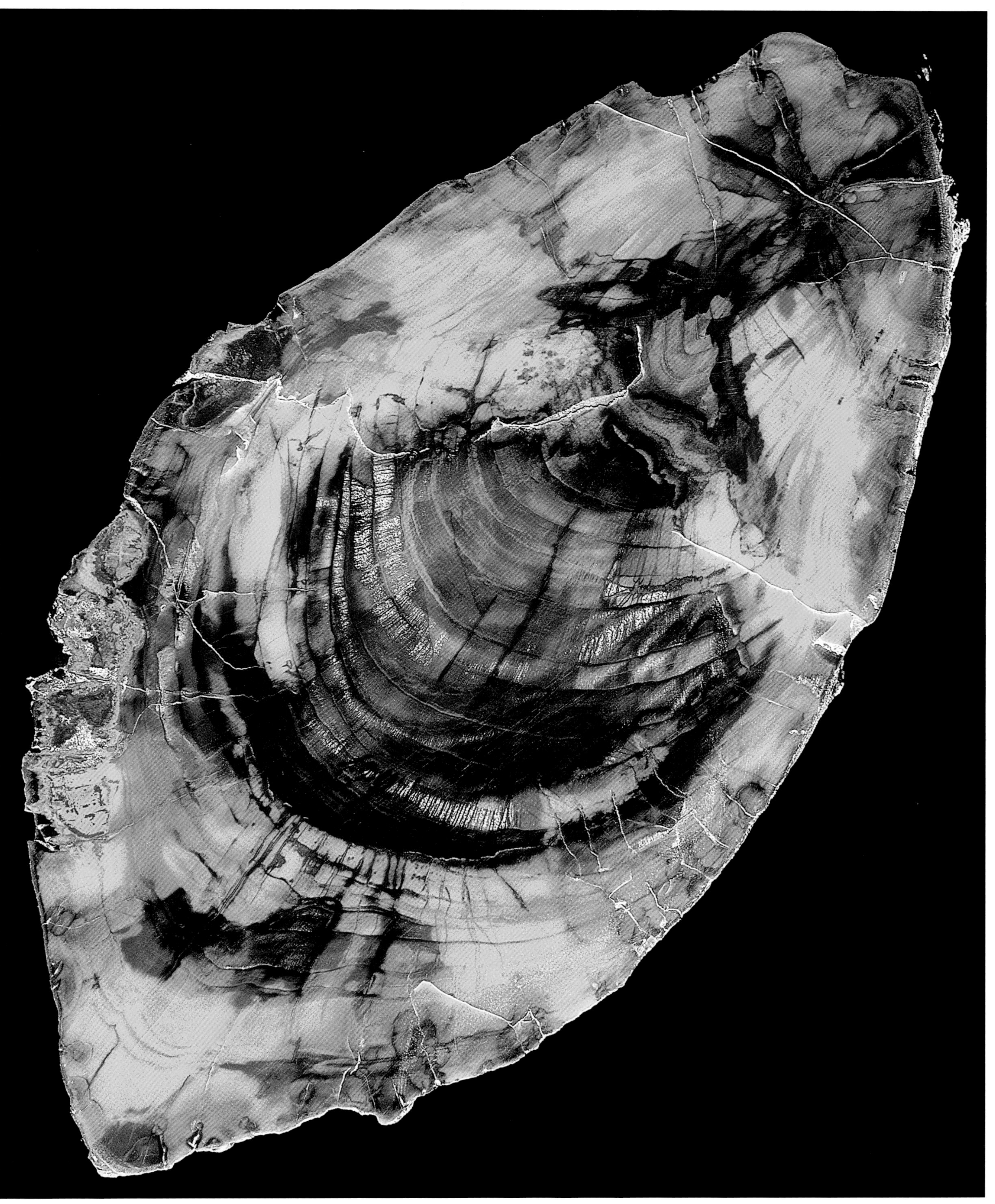

Utah, USA [Henry Mountains]
Mesozoic; Jurassic conifer
fm: Morrison Jones 20 cm

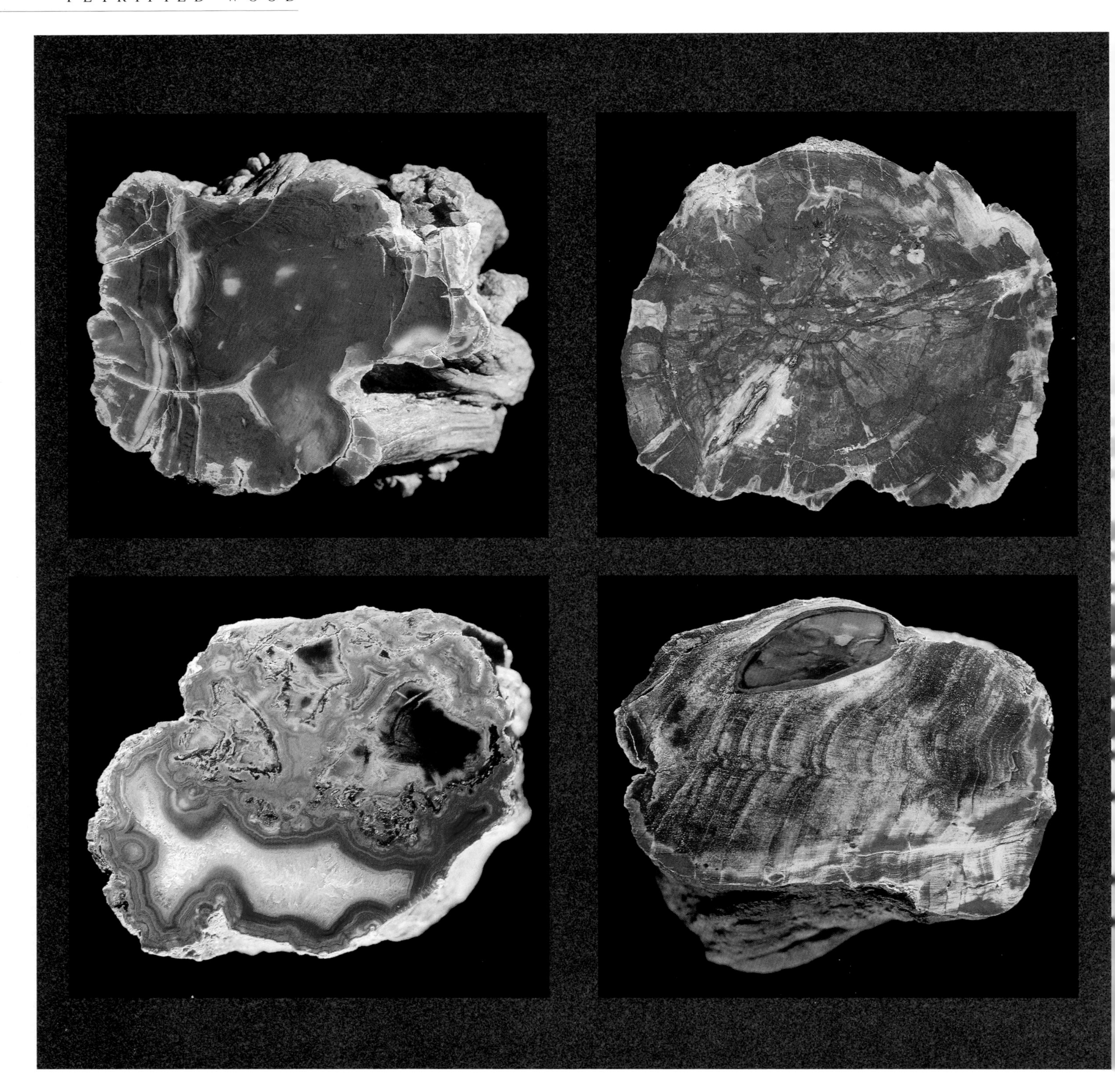

Utah, USA [Henry Mountains]
Mesozoic; Jurassic
unidentified fm: Morrison
Jones 11.5 cm

Utah, USA [Escalante]
Mesozoic; Jurassic
conifer fm: Morrison
Cataldo 13 cm

Utah, USA [Henry Mountains]
Mesozoic; Jurassic
unidentified fm: Morrison
Jones 6.5 cm

Utah, USA [Escalante]
Mesozoic; Jurassic
conifer fm: Morrison
Daniels 7 cm

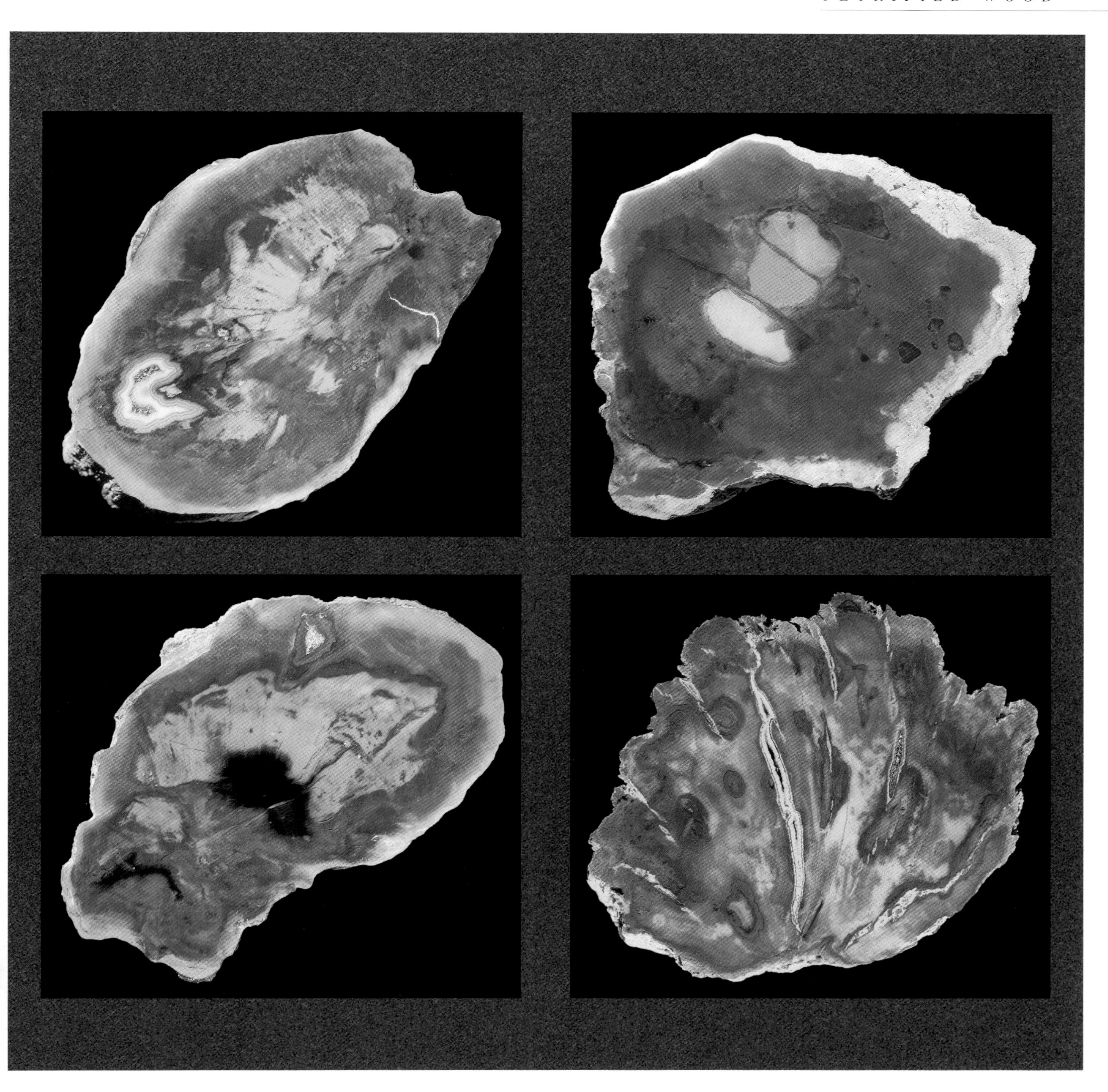

Utah, USA [Henry Mountains]
Mesozoic; Jurassic
conifer fm: Morrison
Jones 11 cm

Utah, USA [Escalante]
Mesozoic; Jurassic
conifer fm: Morrison
Daniels 5 cm

Utah, USA [Henry Mountains]
Mesozoic; Jurassic
conifer fm: Morrison
Jones 13.5 cm

Utah, USA
Mesozoic; Jurassic
unidentified fm: Morrison
Daniels 7 cm

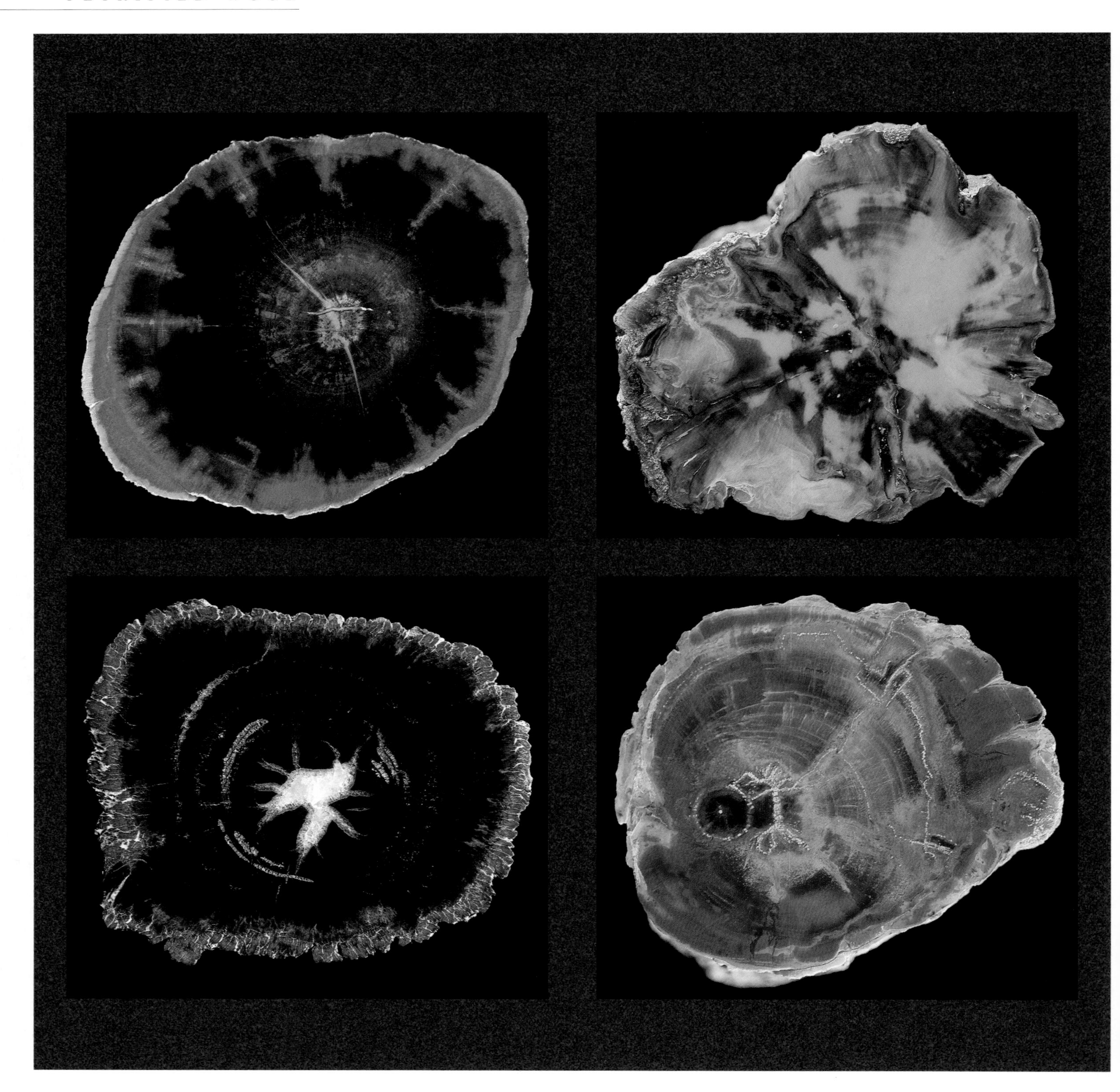

Utah, USA [Circle Cliffs]
Mesozoic; Triassic
conifer fm: Chinle
Daniels 5 cm

Utah, USA [Henry Mountains]
Mesozoic; Jurassic
conifer fm: Morrison
Jones 9.5 cm

Utah, USA [Henry Mountains]
Mesozoic; Jurassic
unidentified fm: Morrison
Jones 10.5 cm

Utah, USA [Henry Mountains]
Mesozoic; Jurassic
conifer fm: Morrison
Daniels 5.5 cm

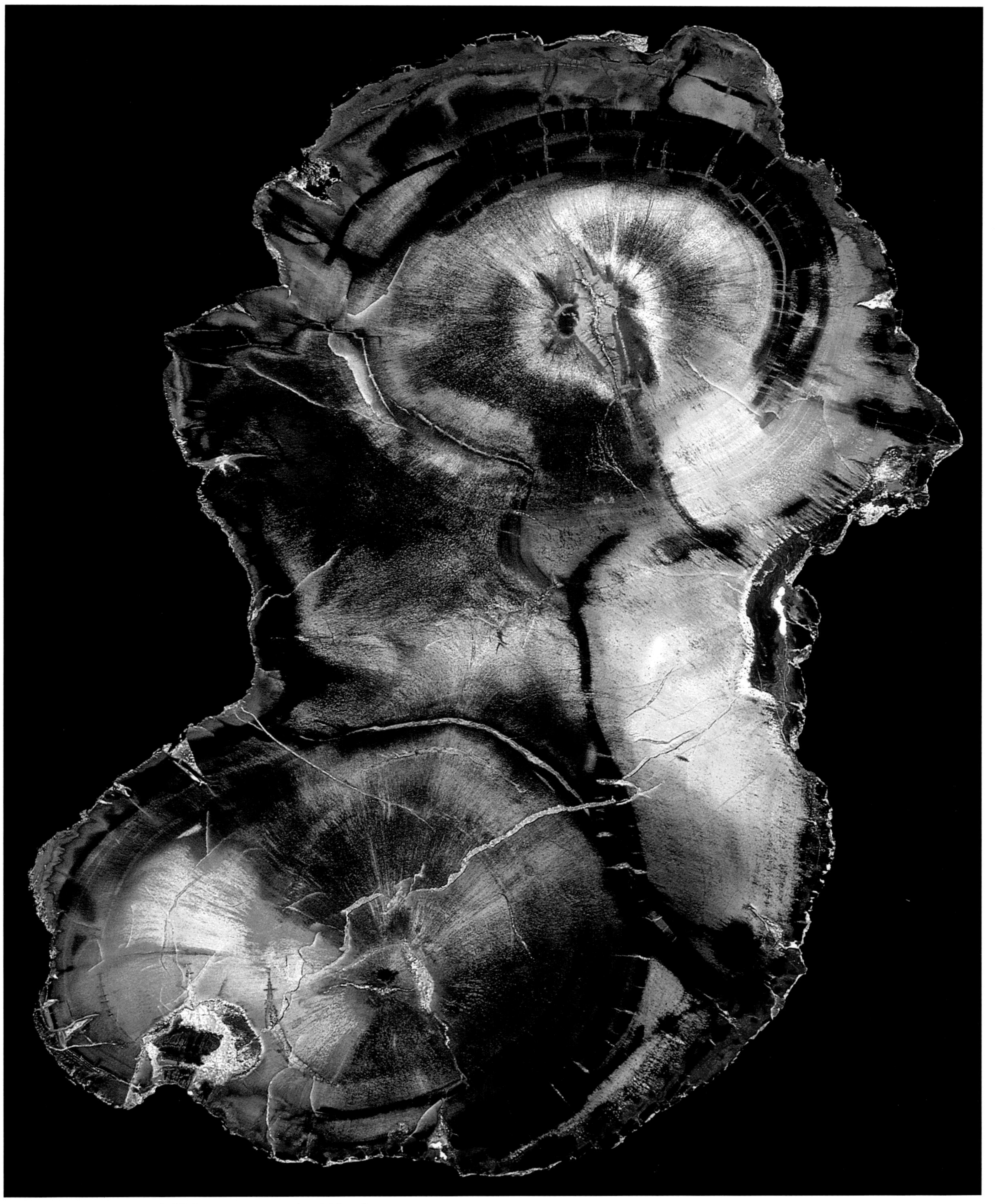

Utah, USA [Circle Cliffs] Mesozoic; Triassic
conifer [double-hearted]
fm: Chinle Daniels 29.5 cm

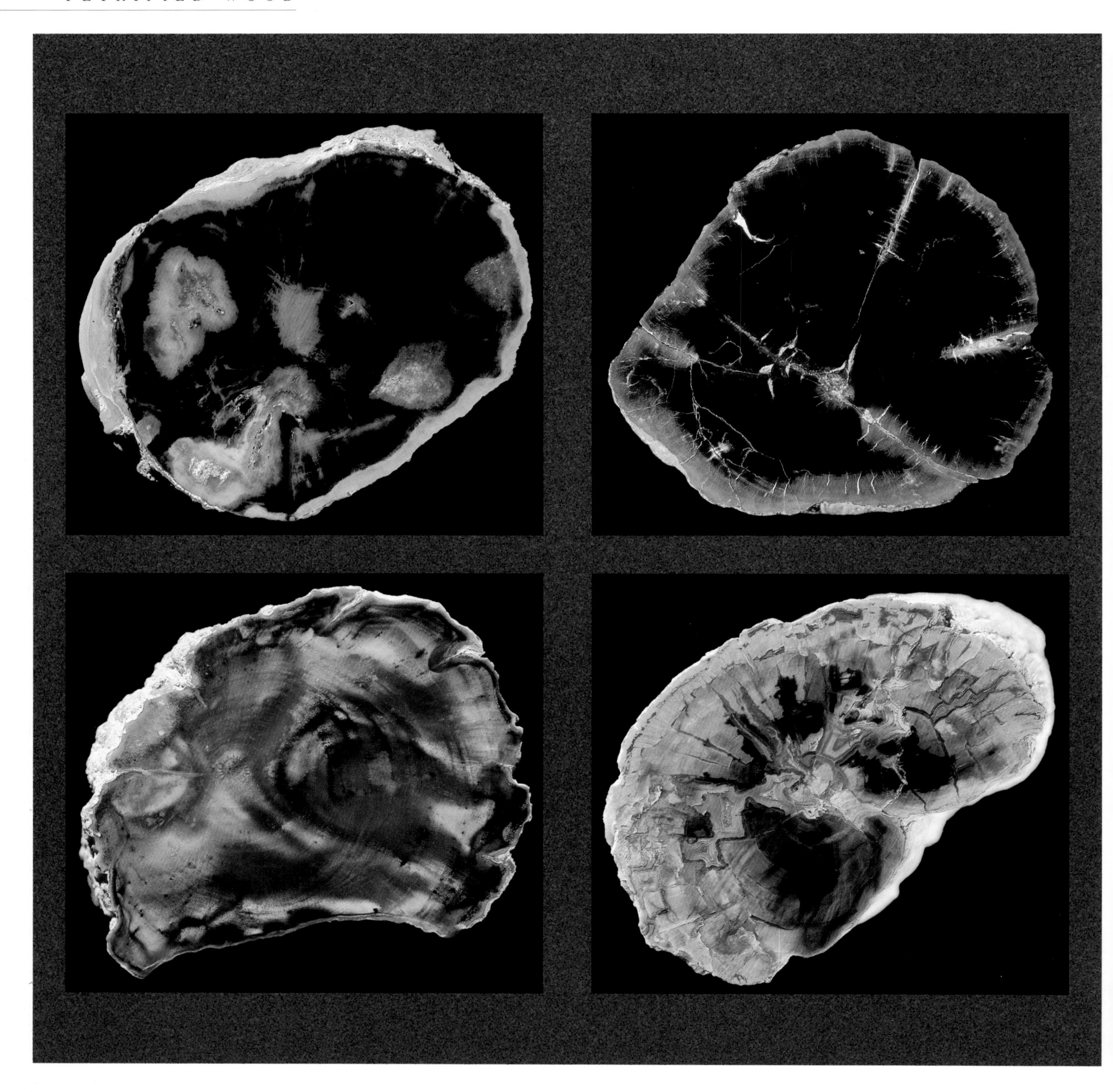

Utah, USA [Henry Mountains]
Mesozoic; Jurassic
conifer fm: Morrison
Daniels 7 cm

Utah, USA [Henry Mountains]
Mesozoic; Jurassic
conifer fm: Morrison
Daniels 8 cm

Utah, USA [Escalante]
Mesozoic; Jurassic
unidentified fm: Morrison
Daniels 5.5 cm

Utah, USA [Henry Mountains]
Mesozoic; Jurassic
conifer fm: Morrison
Jones 11.5 cm

Utah, USA [Circle Cliffs]
Mesozoic; Triassic
conifer fm: Chinle
Daniels 50 cm

Utah, USA
[Henry Mountains]
Mesozoic
conifer
Jones 16 cm

Utah, USA
[Yellow Cat "redwood"]
Mesozoic; Jurassic
unidentified
fm: Morrison
Kladder 11 cm

Utah, USA
[Yellow Cat "redwood"]
Mesozoic; Jurassic
unidentified
fm: Morrison
Dayvault 13 cm

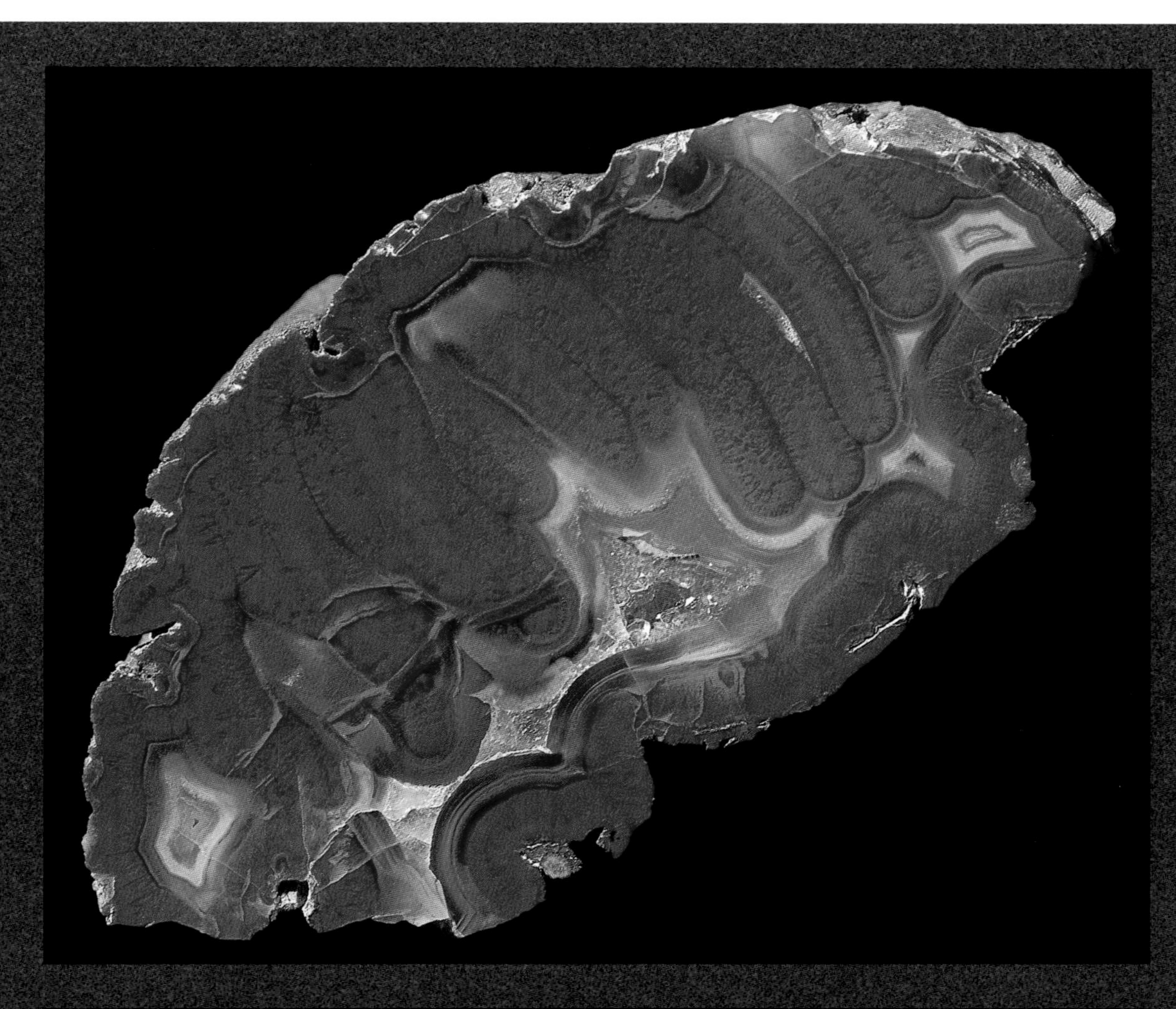

Utah, USA
[Yellow Cat "redwood"]
Mesozoic; Jurassic
unidentified
fm: Morrison
Kladder 9 cm

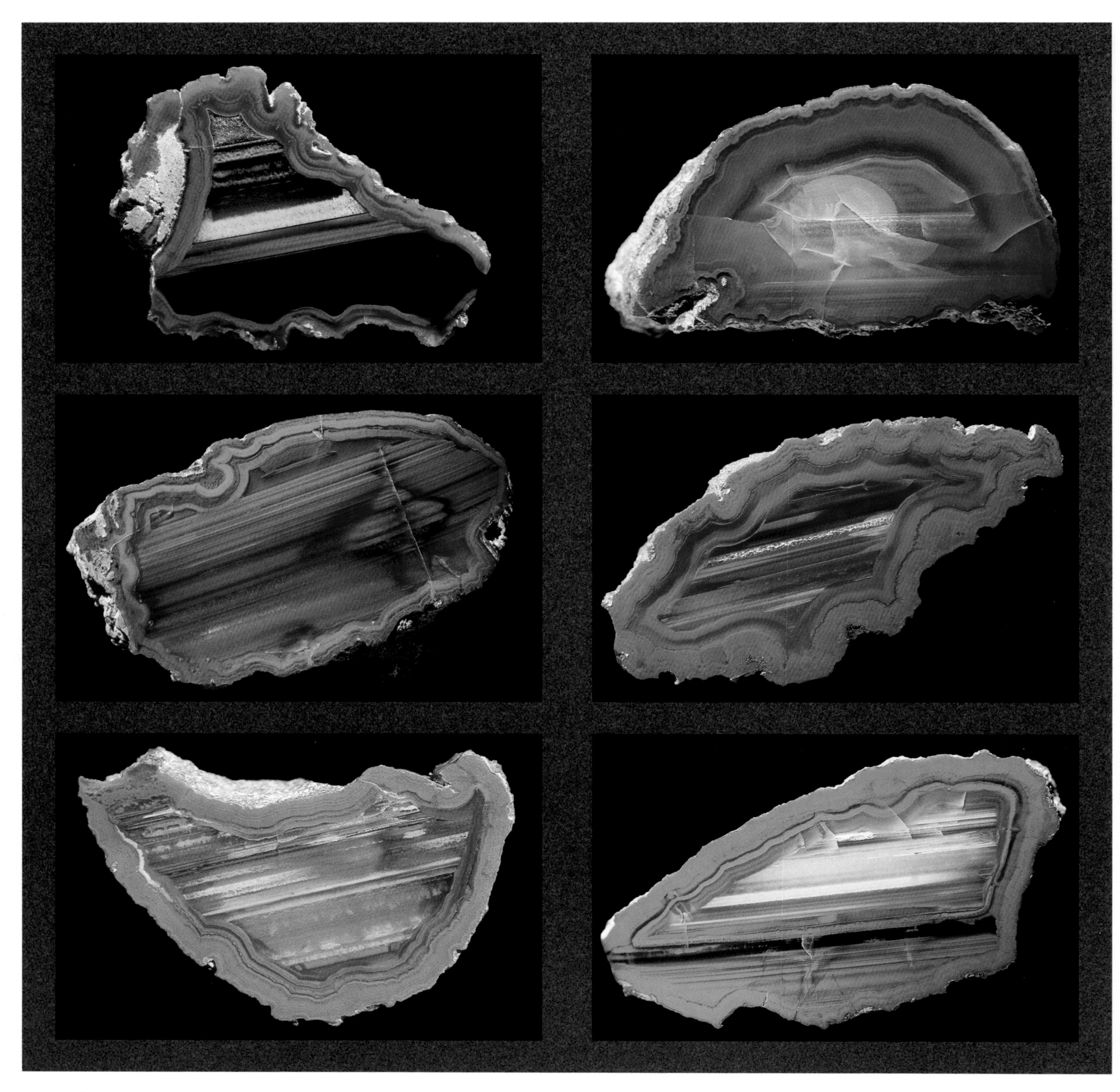

Utah, USA [Yellow Cat "redwood"]
Mesozoic; Jurassic
unidentified fm: Morrison
Kladder 3.5 cm

Utah, USA [Yellow Cat "redwood"]
Mesozoic; Jurassic
unidentified fm: Morrison
Kladder 5 cm

Utah, USA [Yellow Cat "redwood"]
Mesozoic; Jurassic
unidentified fm: Morrison
Kladder 6 cm

Utah, USA [Yellow Cat "redwood"]
Mesozoic; Jurassic
unidentified fm: Morrison
Kladder 4.5 cm

Utah, USA [Yellow Cat "redwood"]
Mesozoic; Jurassic
unidentified fm: Morrison
Kladder 3.5 cm

Utah, USA [Yellow Cat "redwood"]
Mesozoic; Jurassic
unidentified fm: Morrison
Kladder 6 cm

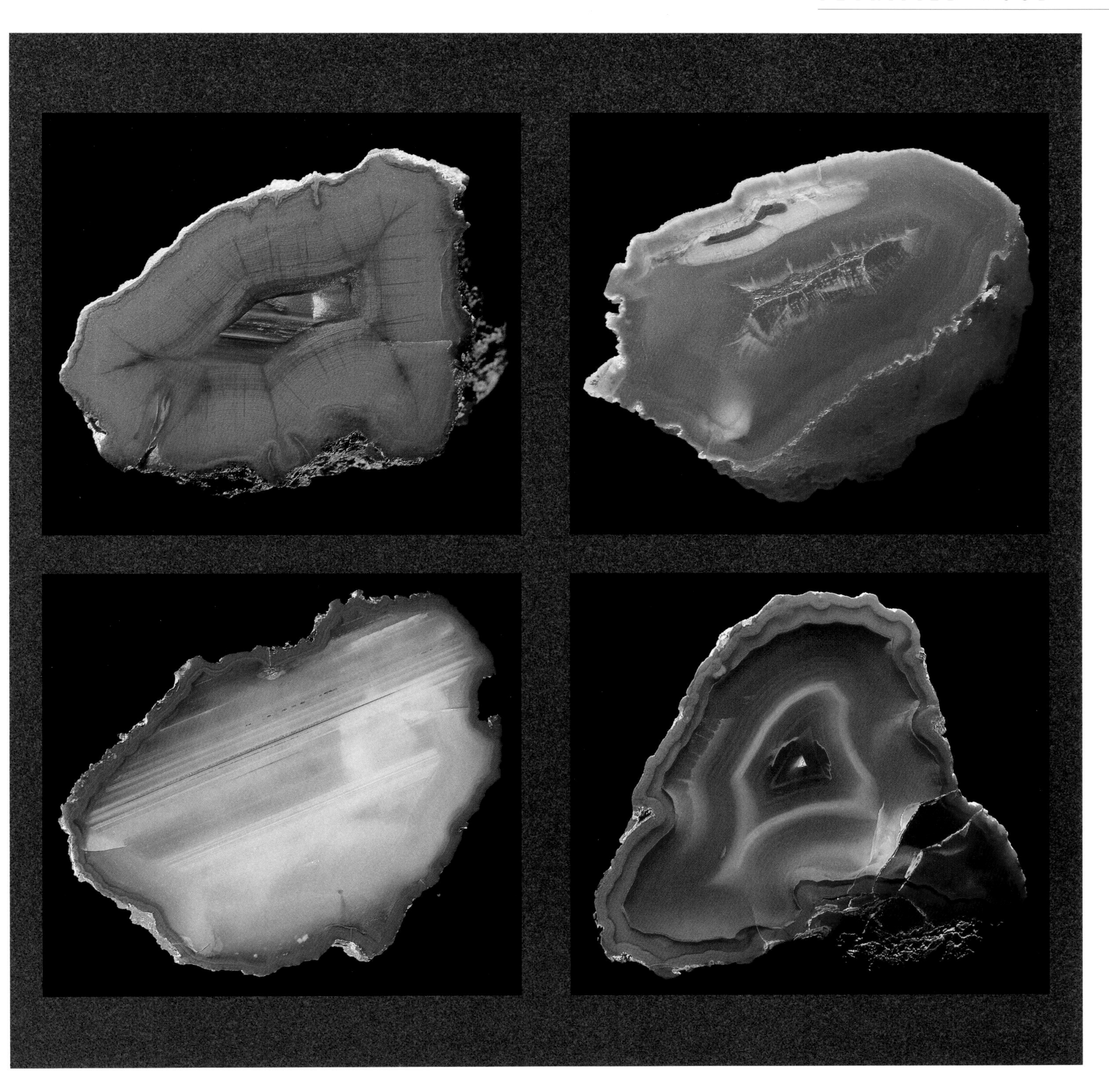

Utah, USA [Yellow Cat "redwood"]
Mesozoic; Jurassic
unidentified fm: Morrison
Kladder 3.5 cm

Utah, USA [Yellow Cat "redwood"]
Mesozoic; Jurassic
unidentified fm: Morrison
Kladder 3.5 cm

Utah, USA [Yellow Cat "redwood"]
Mesozoic; Jurassic
unidentified fm: Morrison
Kladder 6.5 cm

Utah, USA [Yellow Cat "redwood"]
Mesozoic; Jurassic
unidentified fm: Morrison
Kladder 7 cm

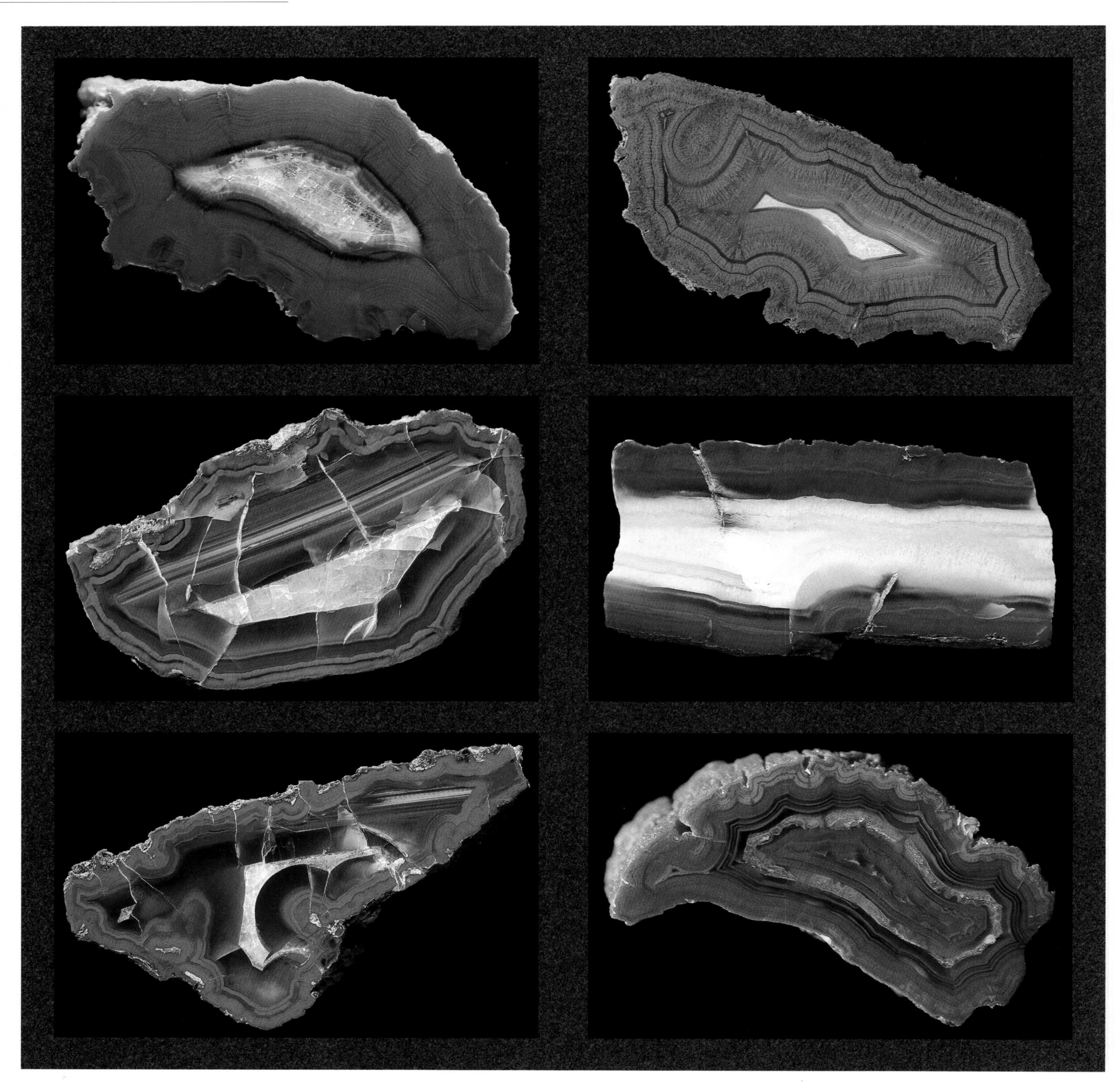

Utah, USA [Yellow Cat "redwood"]
Mesozoic; Jurassic
unidentified fm: Morrison
Kladder 3 cm

Utah, USA [Yellow Cat "redwood"]
Mesozoic; Jurassic
unidentified fm: Morrison
Kladder 9 cm

Utah, USA [Yellow Cat "redwood"]
Mesozoic; Jurassic
unidentified fm: Morrison
Kladder 13 cm

Utah, USA [Yellow Cat "redwood"]
Mesozoic; Jurassic
unidentified fm: Morrison
Kladder 5.5 cm

Utah, USA [Yellow Cat "redwood"]
Mesozoic; Jurassic
unidentified fm: Morrison
Kladder 9 cm

Utah, USA [Yellow Cat "redwood"]
Mesozoic; Jurassic
unidentified fm: Morrison
Kladder 5 cm

Utah, USA
[Yellow Cat "redwood"]
Mesozoic; Jurassic
unidentified
fm: Morrison
Jones 15 cm

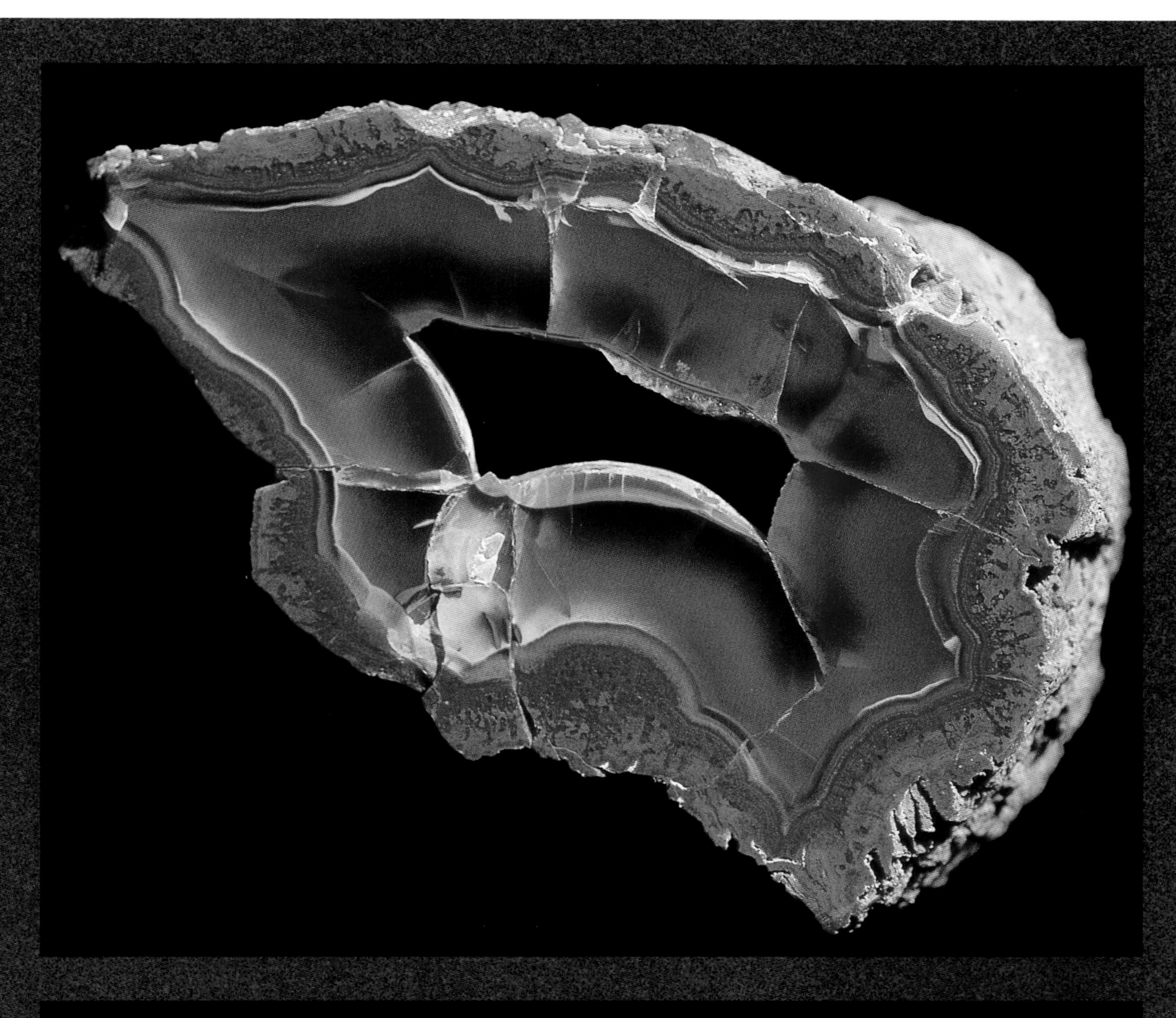

Utah, USA
[Yellow Cat "redwood"]
Mesozoic; Jurassic
unidentified
fm: Morrison
Jones 15 cm

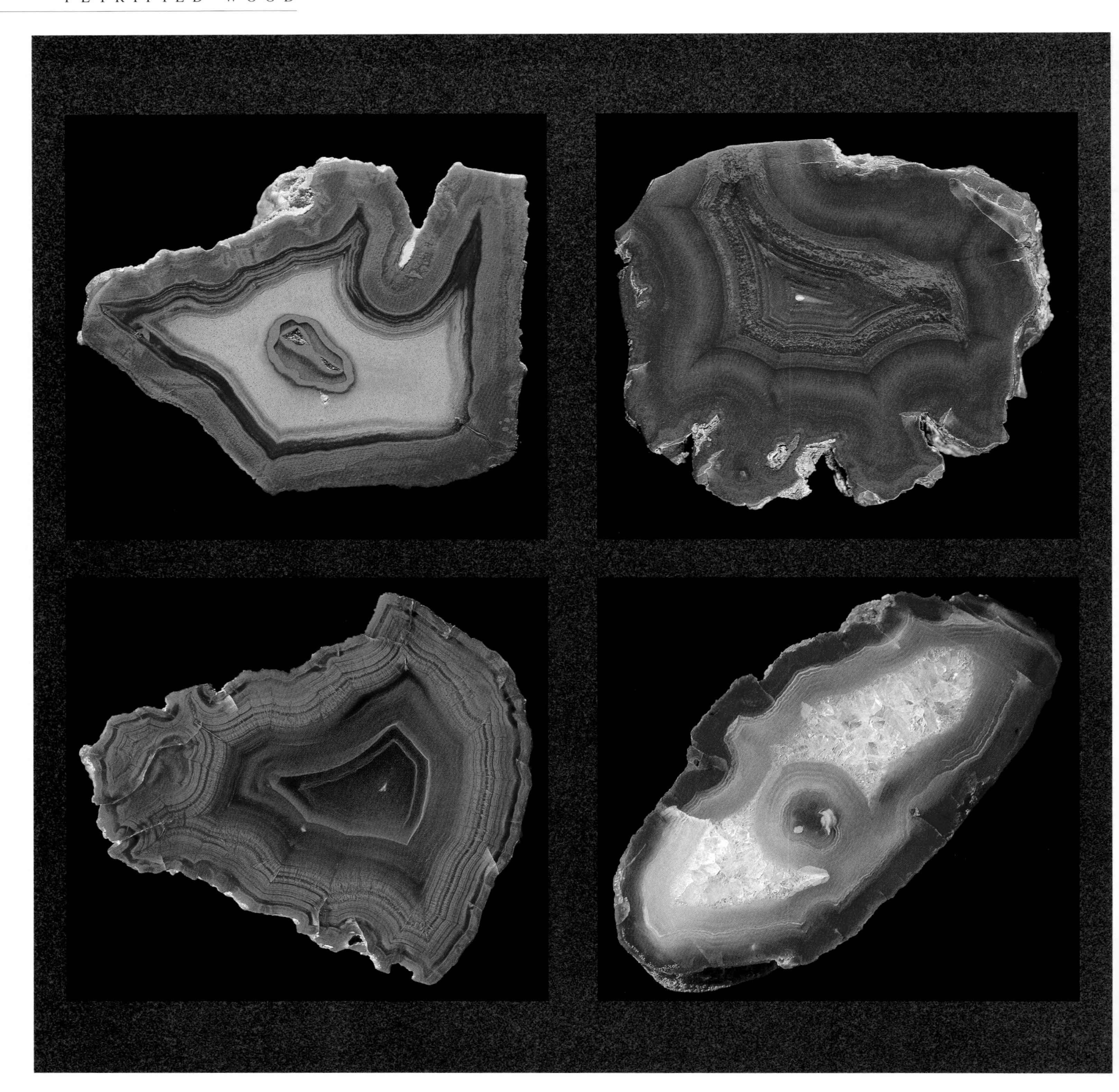

Utah, USA [Yellow Cat "redwood"]
Mesozoic; Jurassic
unidentified fm: Morrison
Jones 5 cm

Utah, USA [Yellow Cat "redwood"]
Mesozoic; Jurassic
unidentified fm: Morrison
Jones 7 cm

Utah, USA [Yellow Cat "redwood]
Mesozoic; Jurassic
unidentified fm: Morrison
Daniels 5 cm

Utah, USA [Yellow Cat "redwood]
Mesozoic; Jurassic
unidentified fm: Morrison
Kladder 9.5 cm

Utah, USA [Yellow Cat "redwood"]
Mesozoic; Jurassic
unidentified fm: Morrison
Kladder 4.5 cm

Utah, USA [Yellow Cat "redwood"]
Mesozoic; Jurassic
unidentified fm: Morrison
Kladder 2.5 cm

Utah, USA [Yellow Cat "redwood"]
Mesozoic; Jurassic
unidentified fm: Morrison
Kladder 4.5 cm

Utah, USA [Yellow Cat "redwood"]
Mesozoic; Jurassic
unidentified fm: Morrison
Kladder 7 cm

WASHINGTON

Washington petrified woods offer many fine examples of permineralization. Tertiary forests and volcanic activity combined to produce a wide variety of petrified woods, preserving trees similar to their surviving descendants. At least thirty-five distinct species of fossil trees have been identified in the Vantage area. Ginkgo trees have remained virtually unchanged for millions of years. The association between volcanism and petrification is quite apparent when petrified logs are dug directly from basalt lava flows in Washington.

Washington, USA [Vantage]
***Ginkgo biloba* [ginkgo]**
Rose

Cenozoic; Tertiary; Miocene
fm: Columbia River Basalt
24 cm

Washington, USA
[Vantage]
Cenozoic; Tertiary;
Miocene
***Alnus* [alder]**
fm: Columbia River Basalt
Rose 16 cm

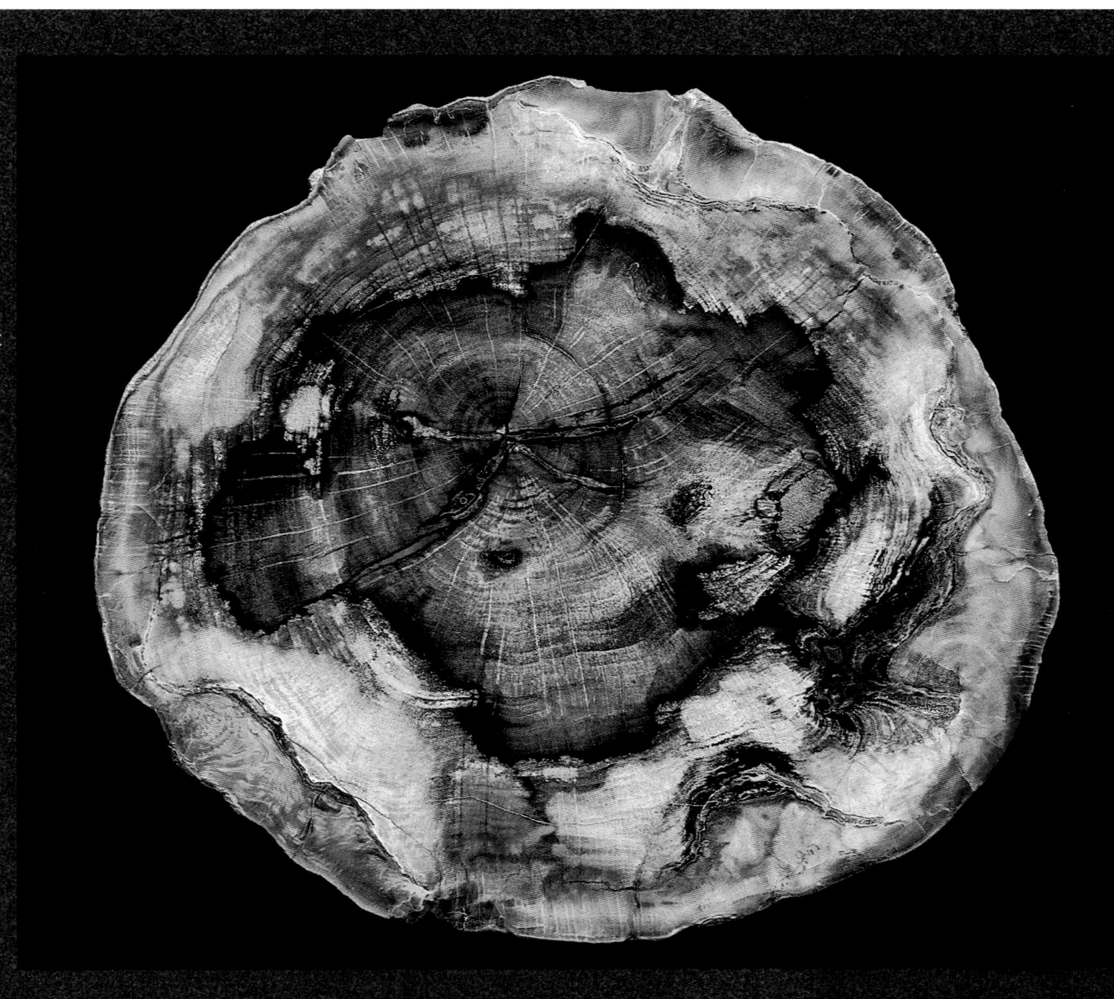

Washington, USA
[Vantage]
Cenozoic; Tertiary;
Miocene
***Ulmus* [elm]**
fm: Columbia River Basalt
Rose 13 cm

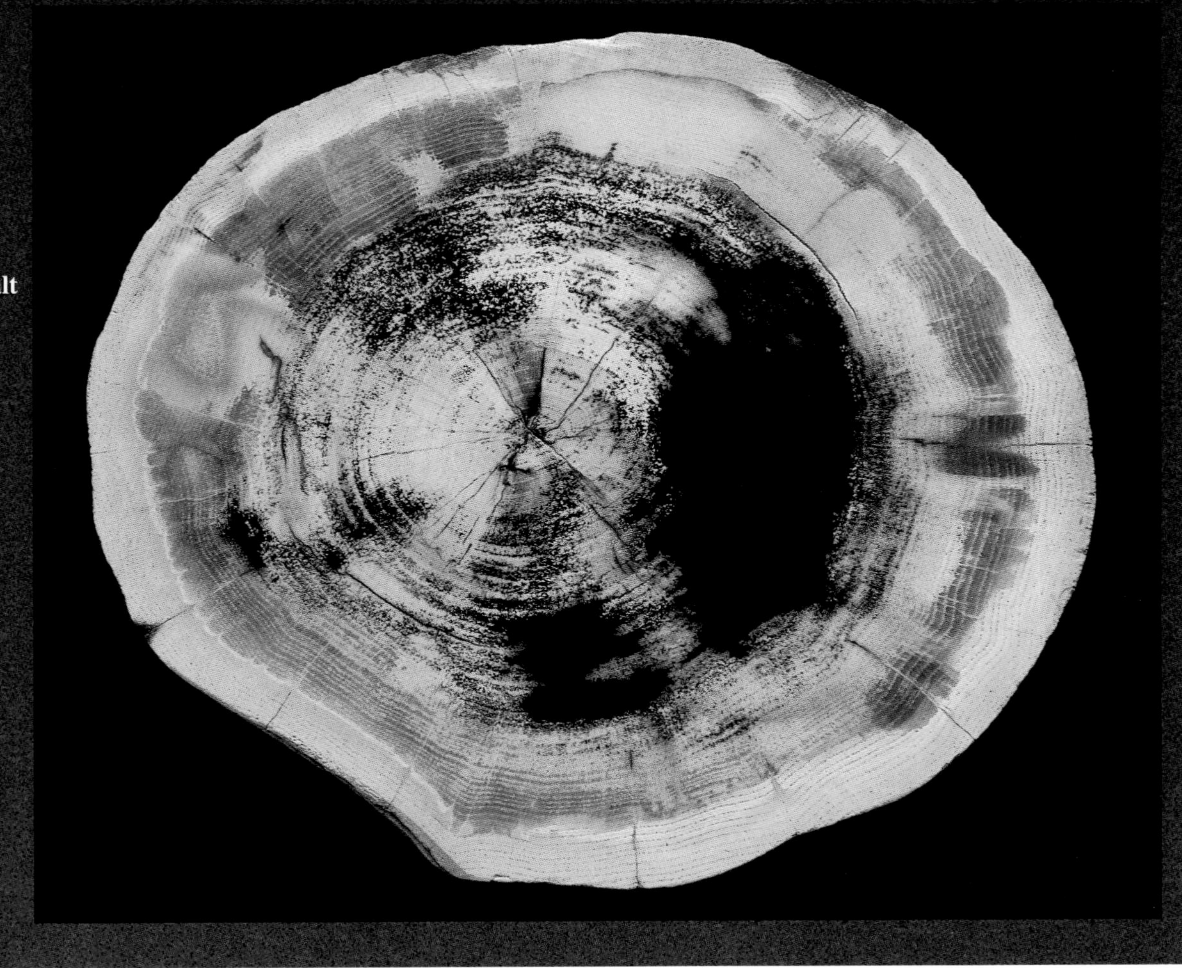

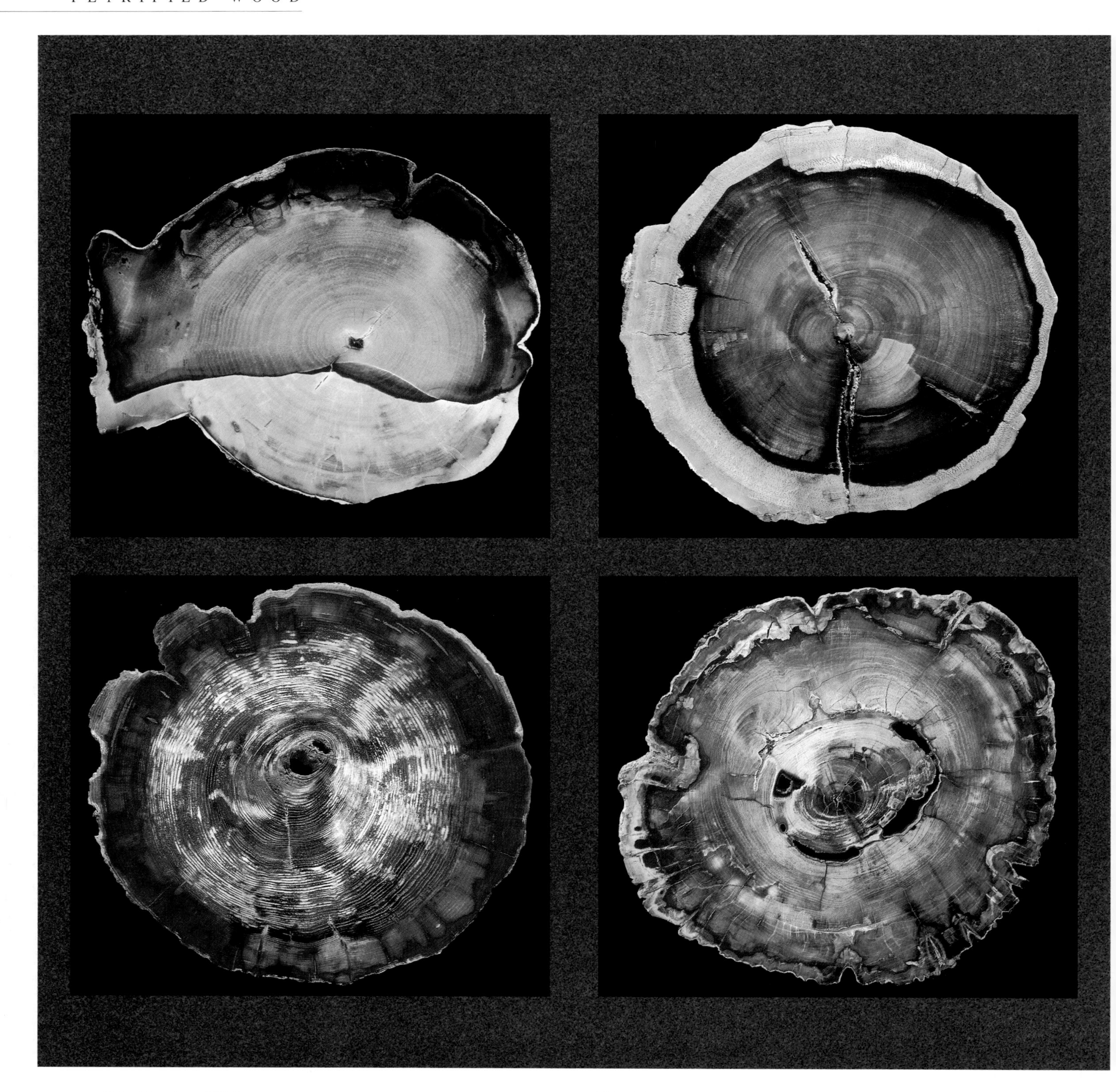

Washington, USA
Cenozoic; Tertiary; Miocene
Sequoia [redwood]
fm: Columbia River Basalt
Daniels 17 cm

Washington, USA [Vantage]
Cenozoic; Tertiary; Miocene
Juglans cinerea [butternut]
fm: Columbia River Basalt
Daniels 15 cm

Washington, USA [Vantage]
Cenozoic; Tertiary; Miocene
Larix [tamarack]
fm: Columbia River Basalt
Rose 21.5 cm

Washington, USA [Vantage]
Cenozoic; Tertiary; Miocene
Acer [maple; fire-scarred]
fm: Columbia River Basalt
Rose 32 cm

Washington, USA
[Vantage]
Cenozoic; Tertiary;
Miocene
***Ginkgo biloba* [ginkgo]**
fm: Columbia River Basalt
Rose 24 cm

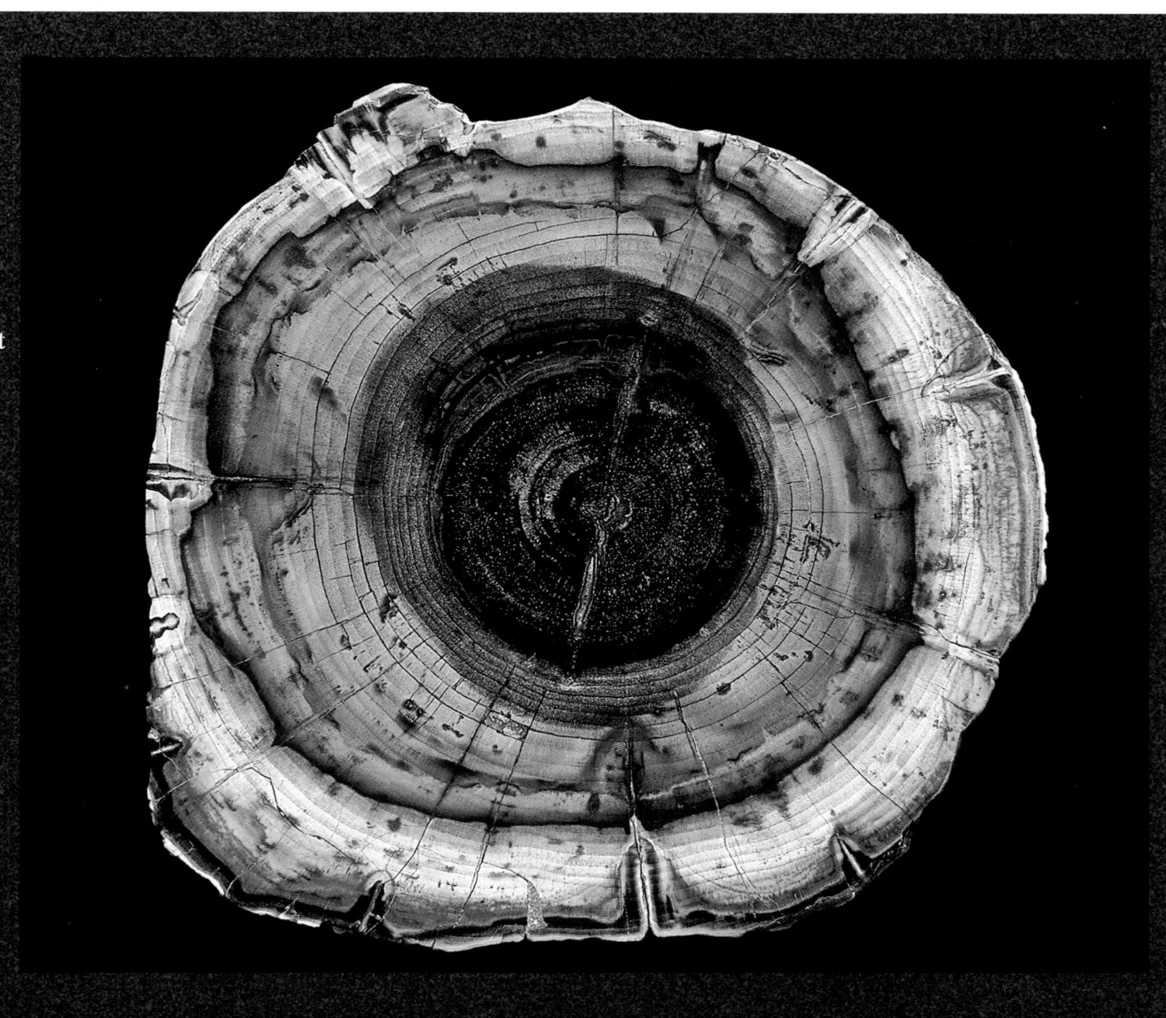

Washington, USA
[Vantage]
Cenozoic; Tertiary;
Miocene
***Ginkgo biloba* [ginkgo]**
fm: Columbia River Basalt
Rose 24 cm

Washington, USA
[Vantage]
Cenozoic; Tertiary;
Miocene
Sequoia [redwood]
fm: Columbia River Basalt
Rose 12 cm

Washington, USA
[Vantage]
Cenozoic; Tertiary;
Miocene
Liquidambar [sweet gum]
fm: Columbia River Basalt
Cataldo 14 cm

Washington, USA [Vantage]
Cenozoic; Tertiary; Miocene
***Pseudotsuga* [Douglas fir]**
fm: Columbia River Basalt
Rose 24 cm

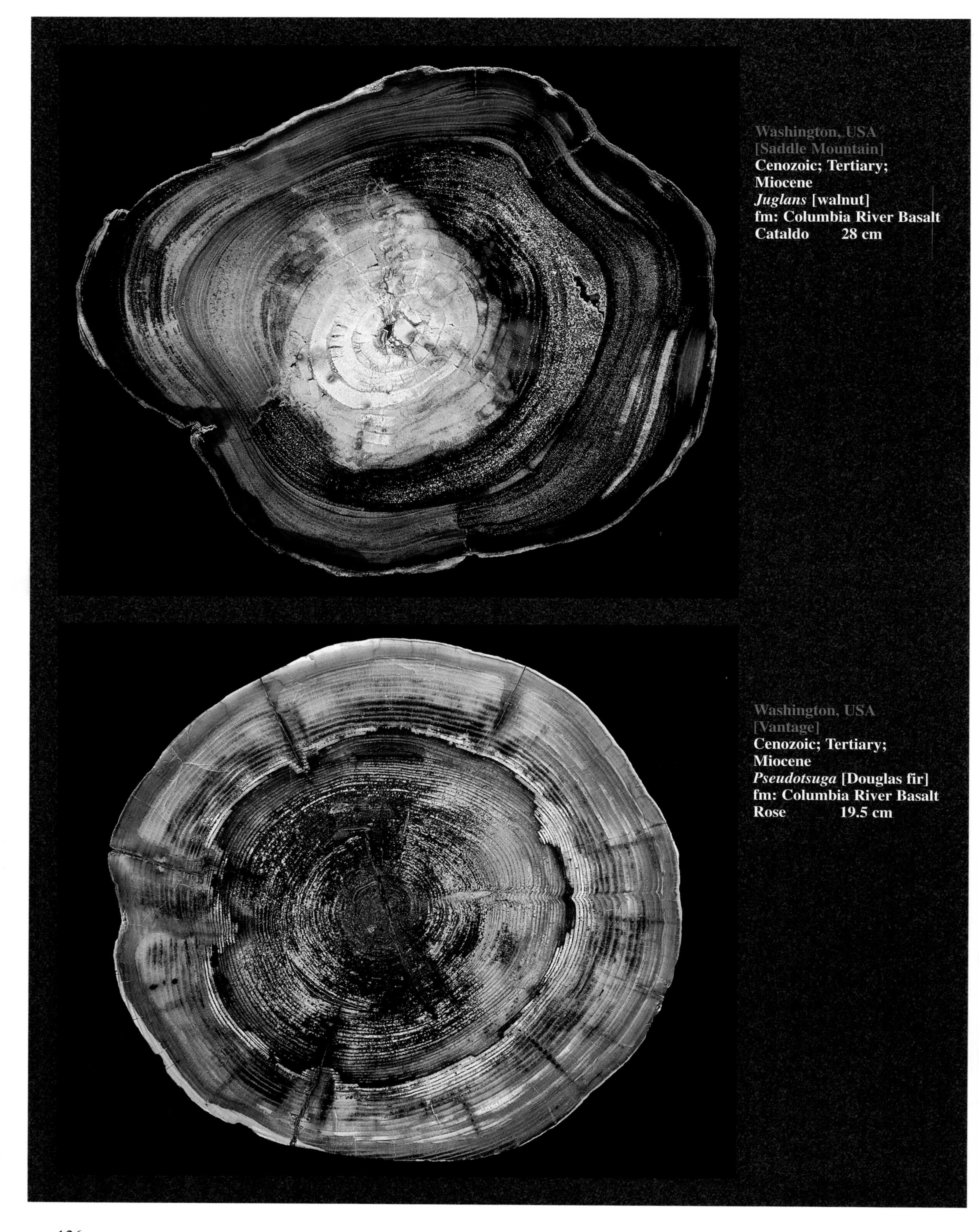

Washington, USA
[Saddle Mountain]
Cenozoic; Tertiary;
Miocene
***Juglans* [walnut]**
fm: Columbia River Basalt
Cataldo 28 cm

Washington, USA
[Vantage]
Cenozoic; Tertiary;
Miocene
***Pseudotsuga* [Douglas fir]**
fm: Columbia River Basalt
Rose 19.5 cm

WYOMING

One of the petrified logs from Wyoming shown in these photographs demonstrates how a single log can transform entirely to quartz crystals at one point and to brown permineralized wood in a jacket of blue chalcedony at another. The state offers a wide variety of petrified wood, including palm and hardwoods. Eden Valley wood is known for its exterior, which resembles a piece of fallen, weathered wood. Blue Forest wood is a unique example of the various forms of petrification. Paleoentomologists find an abundance of materials for examination in the insect-damaged petrifications of this state.

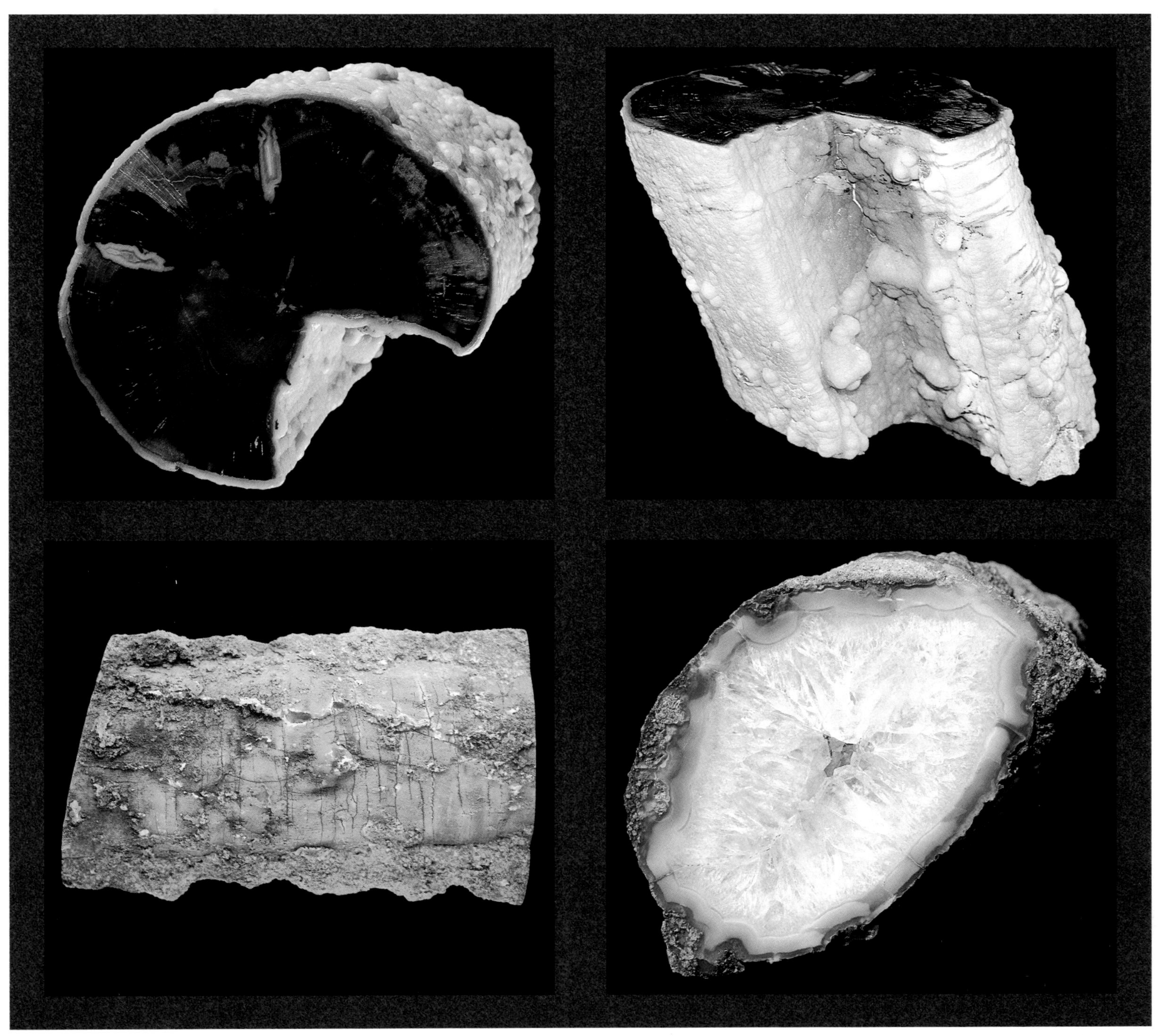

The two specimens pictured are part of the same tree as discovered in the Blue Forest area some years ago; one has been completely replaced by quartz crystals; the other is permineralized.

Wyoming, USA [Blue Forest]
Cenozoic; Tertiary; Eocene
unidentified fm: Green River
Jones *UPPER:* **8.5 cm tall by 8.5 cm**
LOWER: **16.5 cm long by 10 cm**

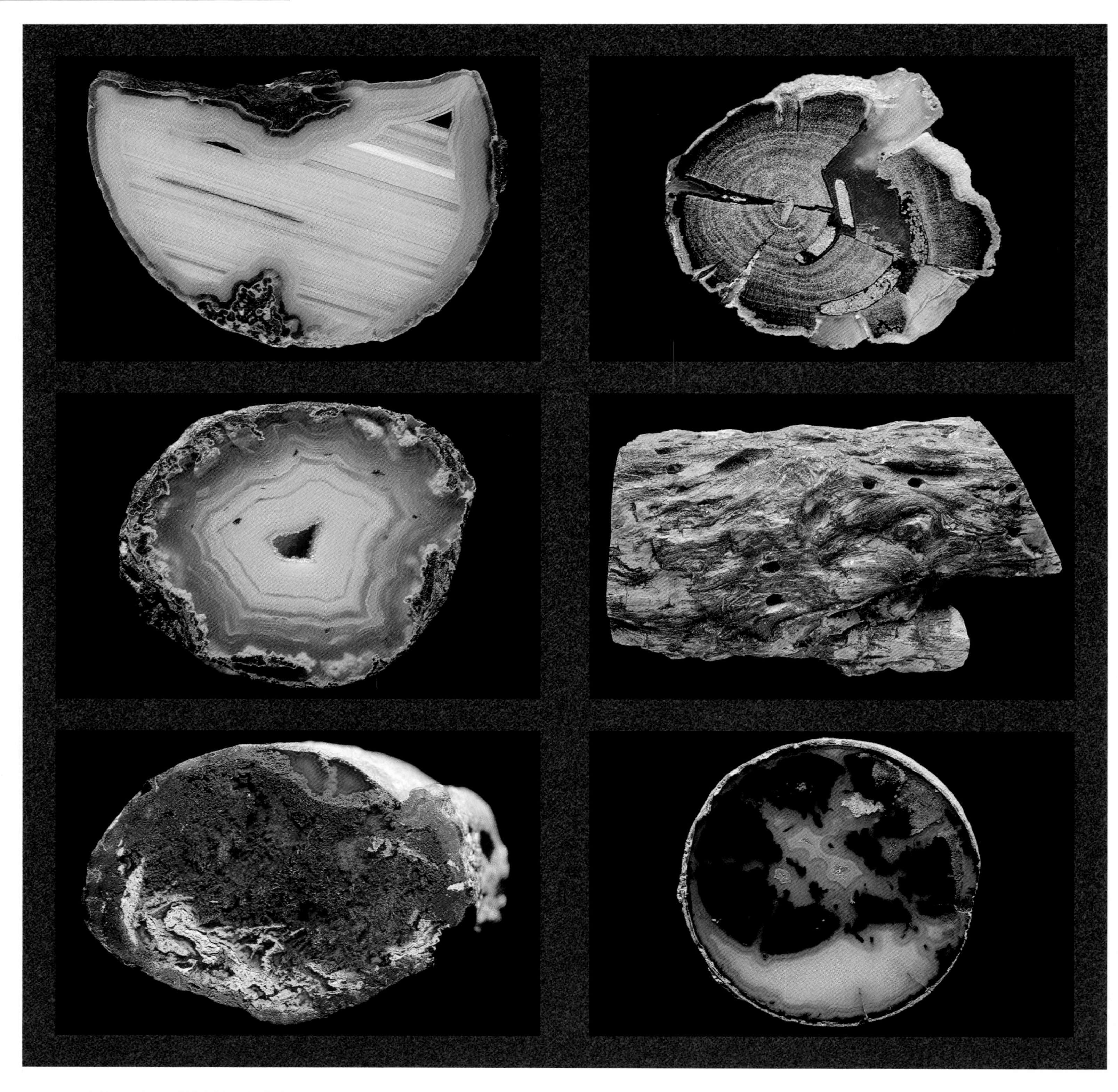

Wyoming, USA [Dubois]
Cenozoic; Tertiary; Eocene
unidentified
Jones 11 cm

Wyoming, USA [Dubois]
Cenozoic; Tertiary; Eocene
unidentified
Jones 4.5 cm

Wyoming, USA [Dubois]
Cenozoic; Tertiary; Eocene
unidentified
Jones 7 cm

Wyoming, USA [Eden Valley]
Cenozoic; Tertiary; Eocene
unidentified [with insect borings]
fm: Green River Jones 4 cm

Wyoming, USA
Cenozoic; Tertiary
unidentified [with insect borings]
Jones 10.5 cm

Wyoming, USA [Farson]
Cenozoic; Tertiary; Eocene
Palmoxylon fm: Green River
Daniels 4.5 cm

Wyoming, USA
[Blue Forest]
Cenozoic; Tertiary;
Eocene
unidentified
fm: Green River
Jones 11.5 cm

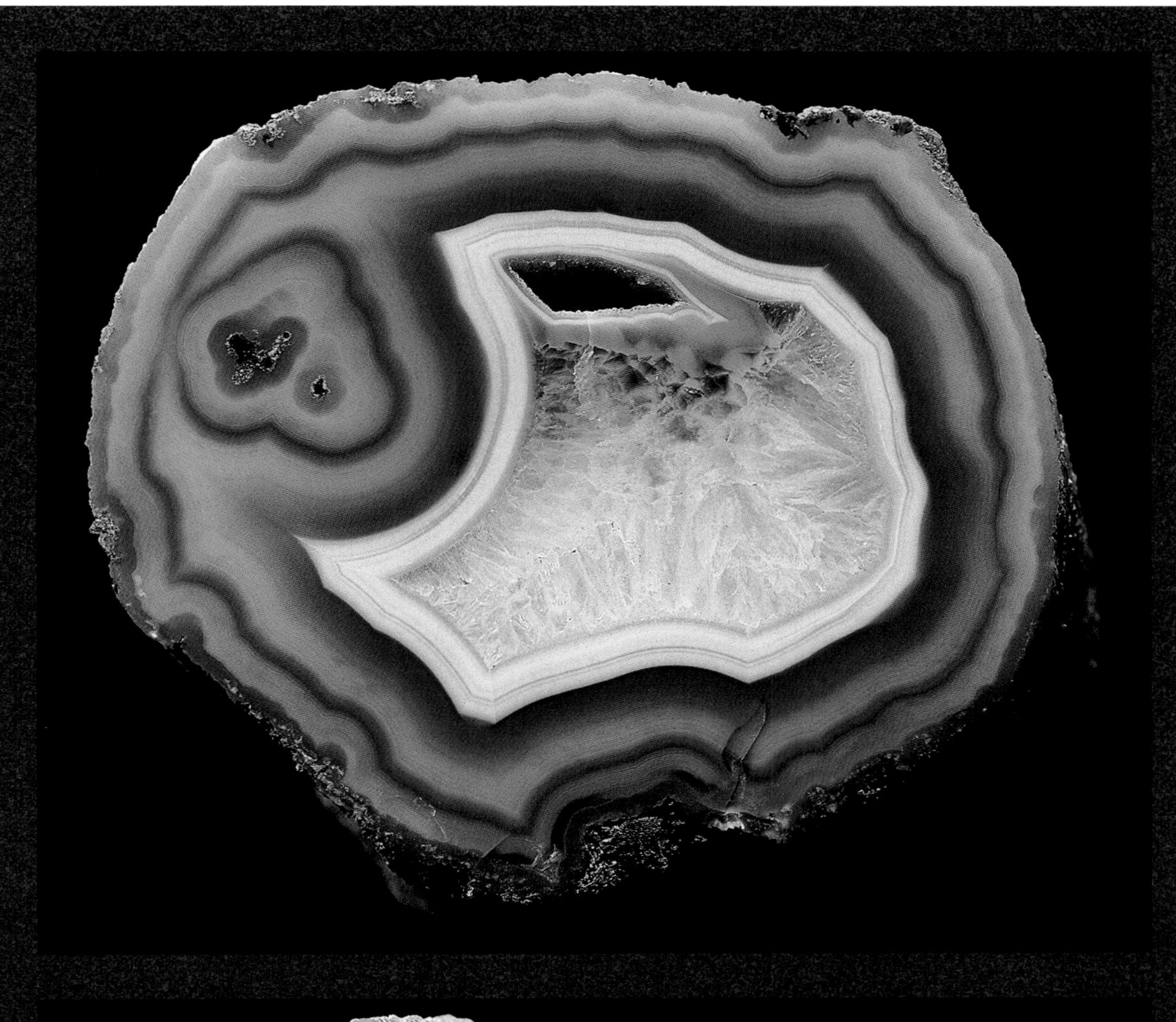

Wyoming, USA [Dubois]
Cenozoic; Tertiary;
Eocene
unidentified
Jones 12 cm

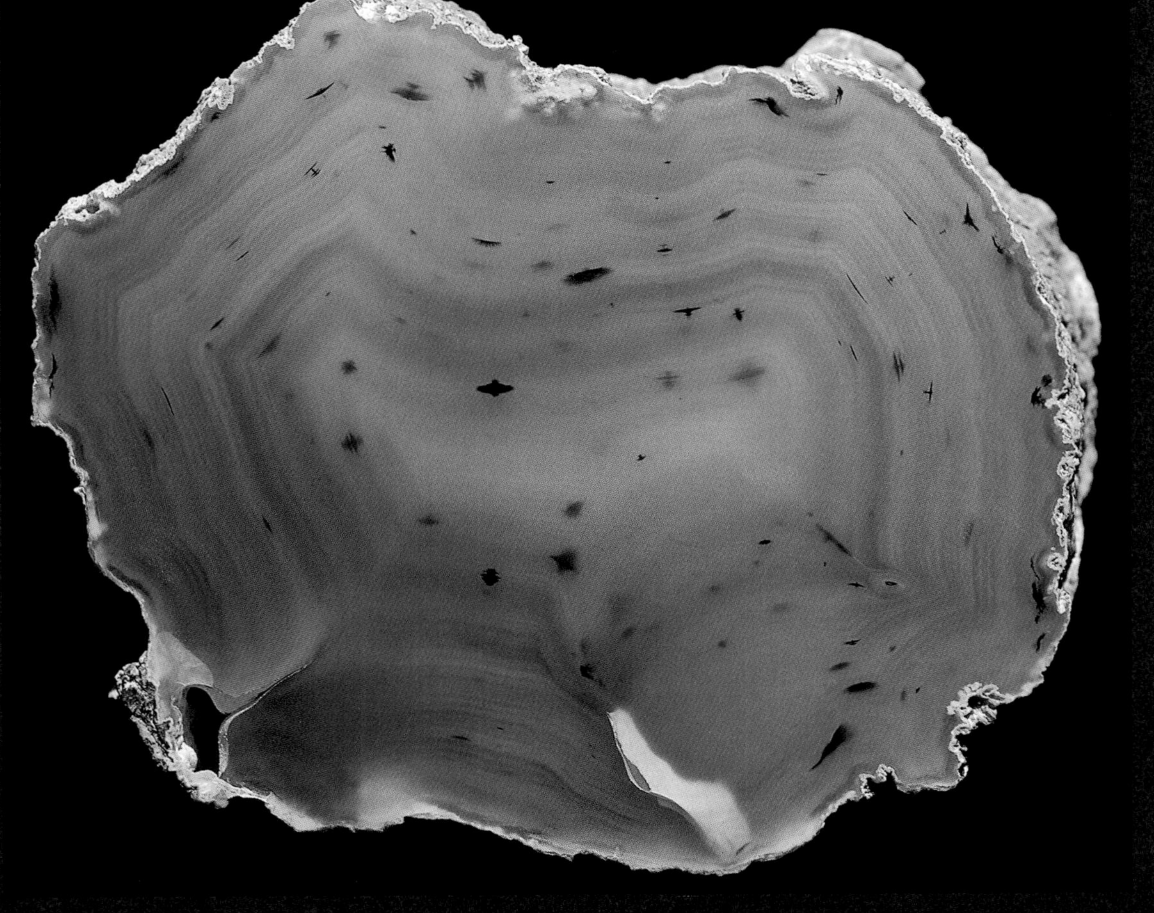

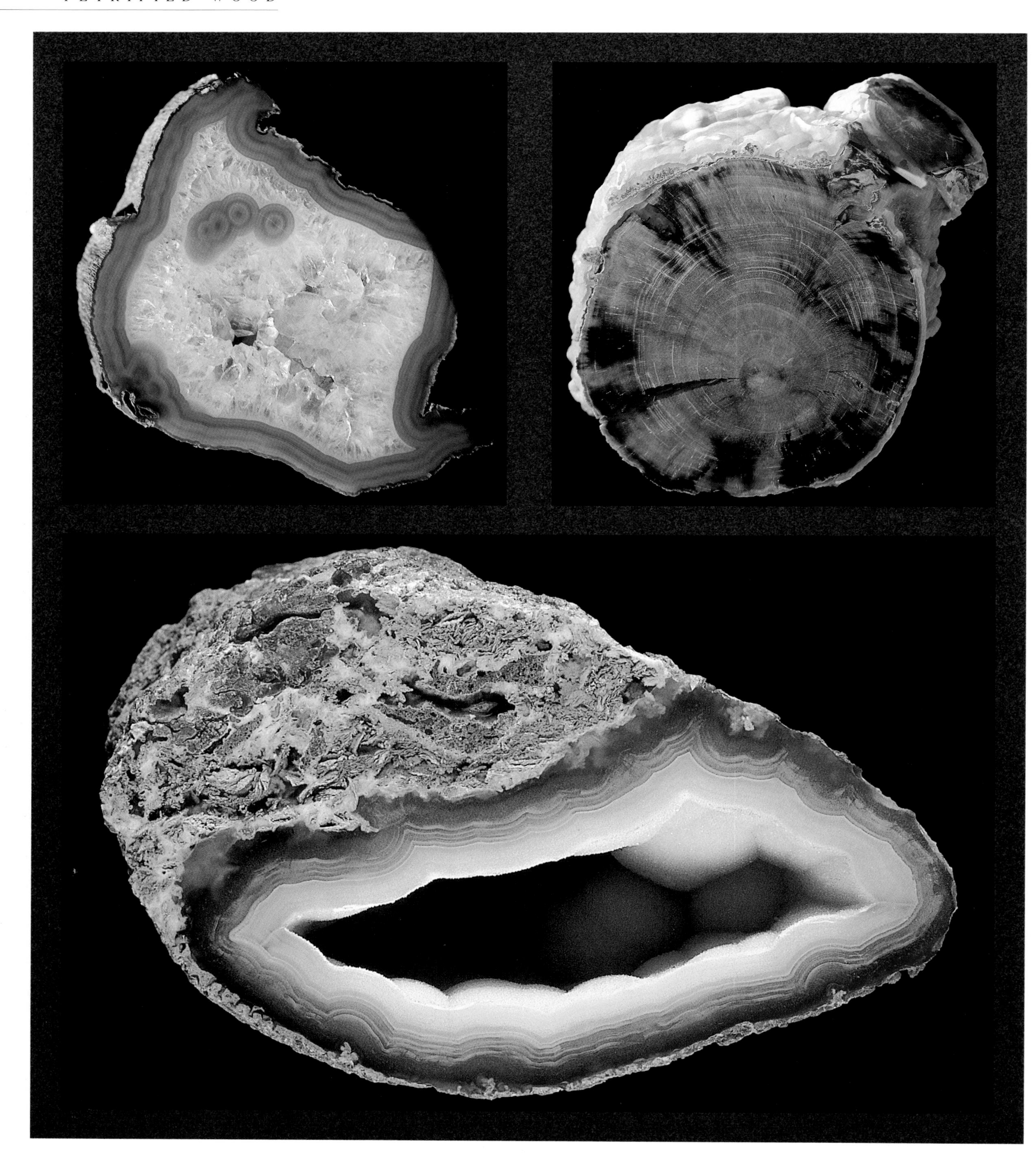

(UPPER LEFT)
Wyoming, USA
Cenozoic; Tertiary; Eocene
unidentified
Daniels 11.5 cm

(BOTTOM)
Wyoming, USA
Cenozoic; Tertiary
unidentified
Jones 11 cm

(UPPER RIGHT)
Wyoming, USA [Blue Forest]
Cenozoic; Tertiary; Eocene
unidentified fm: Green River
Jones small branch 3.5 cm
large branch 11 cm

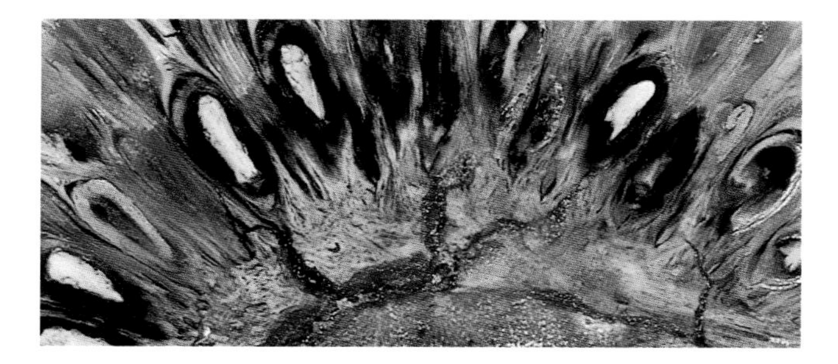

6 Cones

"It is amazing to note the utter rarity of petrified cones."
Professor G.R. Wieland, 1935

Cone-bearing plants encase seeds in cones rather than having them embodied in fruits as is the case with the more recently developed flowering plants. Coniferous trees and other cone-bearing gymnosperms come in a variety of shapes and sizes. There are cycad cones, hemlock cones, spruce cones, *Sequoia* cones, *Araucaria* cones, *Pararaucaria* cones, fir cones, and others in addition to pine cones. The majority of cones pictured in this volume are *Araucaria* cones from Patagonia, Argentina. *Pararaucaria* are attributed to the family Taxodiaceae rather than Araucariaceae, despite the name.

While events and circumstances necessary for the petrification of wood occurred only rarely, they far more rarely combined to result in the petrification of cones. Wood is wood throughout the year; a complete fossil cone had to have begun the petrification process during a very limited stage of its development. Usually only unopened yet somewhat mature cones tightly holding seeds could completely petrify. The small cones often had to survive cataclysmic events such as floods and volcanic flows intact and yet still become buried in an environment suitable for silicification. Fortunately for admirers of petrified cones, the conditions in Jurassic Argentina were ideal. These trees with nearly mature cones were buried in a thick blanket of volcanic ash. Often these cones come to us as completely silicified whole specimens.

Patagonia, Argentina
Mesozoic; Jurassic
Araucaria
Daniels

United States of America
Cretaceous to Tertiary
***Sequoia* [redwood]**
Daniels

Patagonia, Argentina
Mesozoic; Jurassic
Araucaria mirabilis
Daniels 7.5 cm

Patagonia, Argentina
Mesozoic; Jurassic
Araucaria mirabilis
Daniels 6.5 cm

Patagonia, Argentina
Mesozoic; Jurassic
Araucaria mirabilis
Daniels 9.5 cm

Patagonia, Argentina
Mesozoic; Jurassic
Araucaria mirabilis
Daniels 8 cm

Patagonia, Argentina
Mesozoic; Jurassic
Araucaria mirabilis
Daniels 6 cm

Patagonia, Argentina
Mesozoic; Jurassic
Araucaria mirabilis
Daniels 6.5 cm

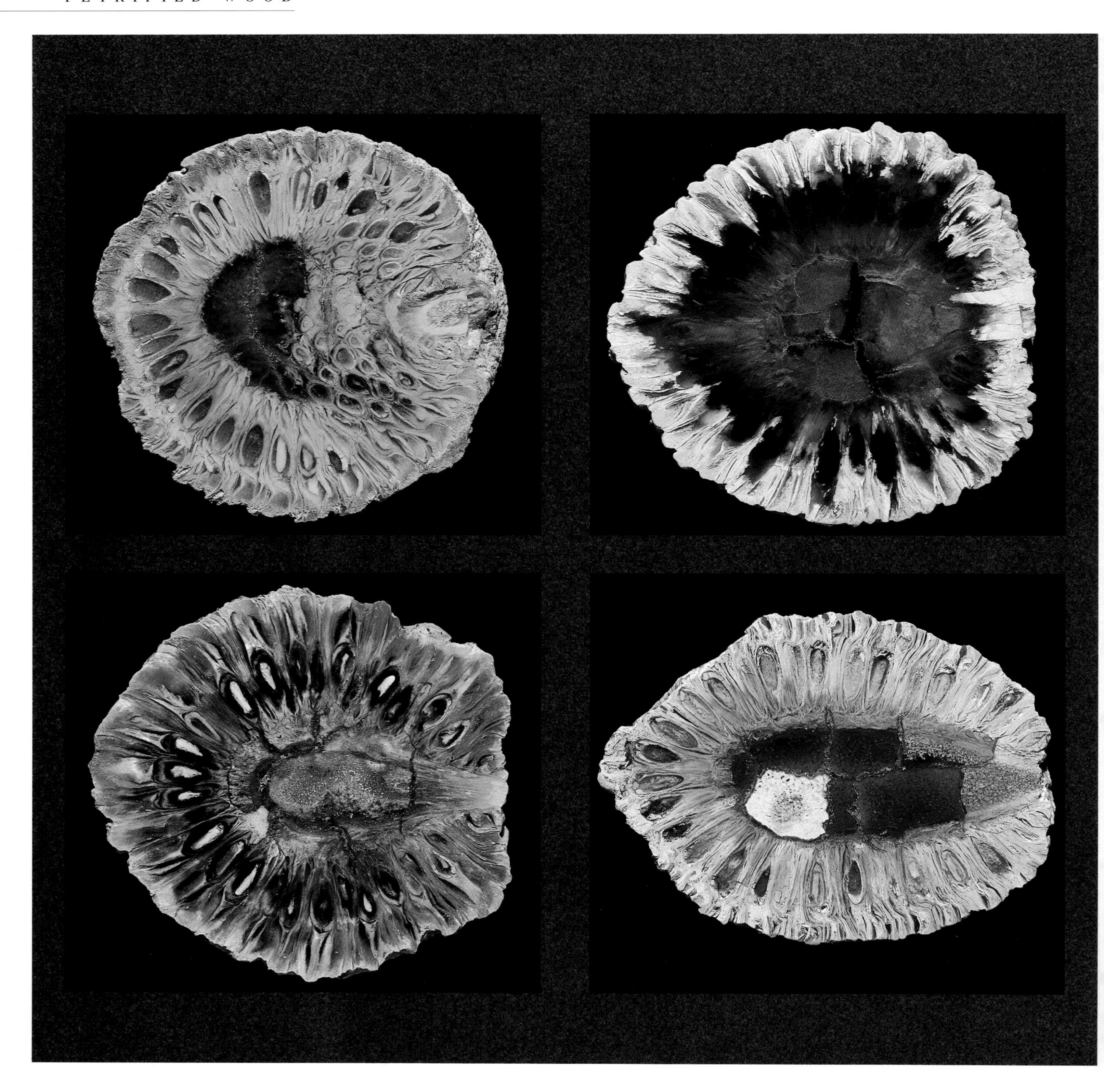

Patagonia, Argentina
Mesozoic; Jurassic
Araucaria mirabilis
Daniels 6 cm

Patagonia, Argentina
Mesozoic; Jurassic
Araucaria mirabilis
Daniels 7 cm

Patagonia, Argentina
Mesozoic; Jurassic
Araucaria mirabilis
Daniels 6 cm

Patagonia, Argentina
Mesozoic; Jurassic
Araucaria mirabilis
Daniels 7 cm

Patagonia, Argentina
Mesozoic; Jurassic
Araucaria
Daniels 8 cm

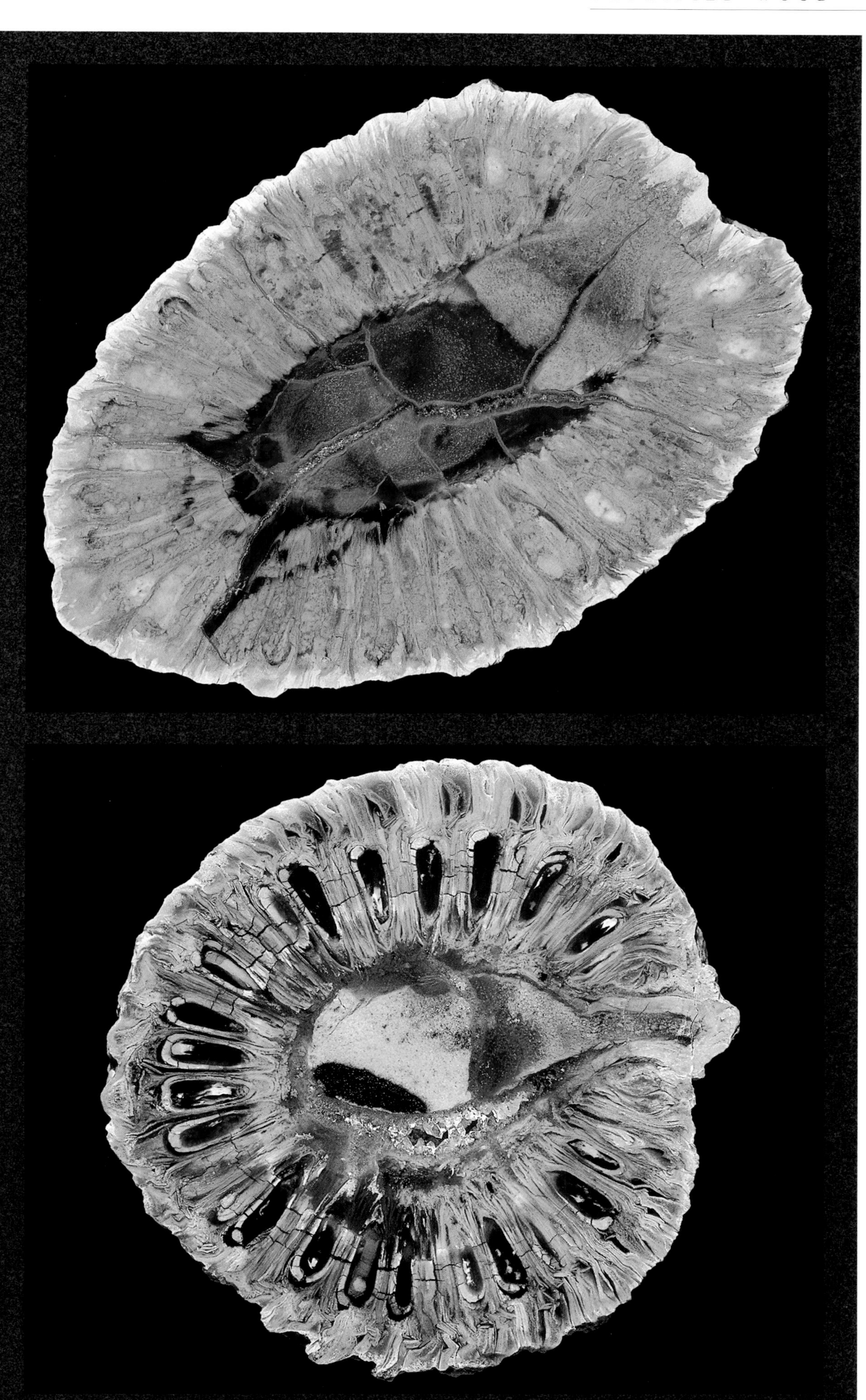

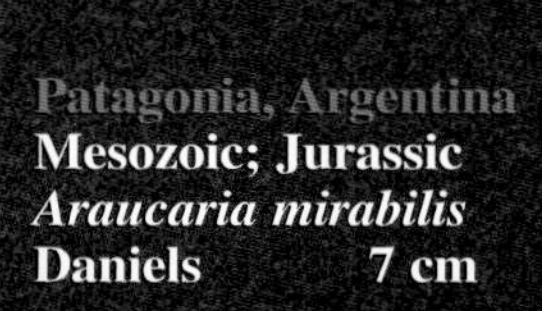

Patagonia, Argentina
Mesozoic; Jurassic
Araucaria mirabilis
Daniels 7 cm

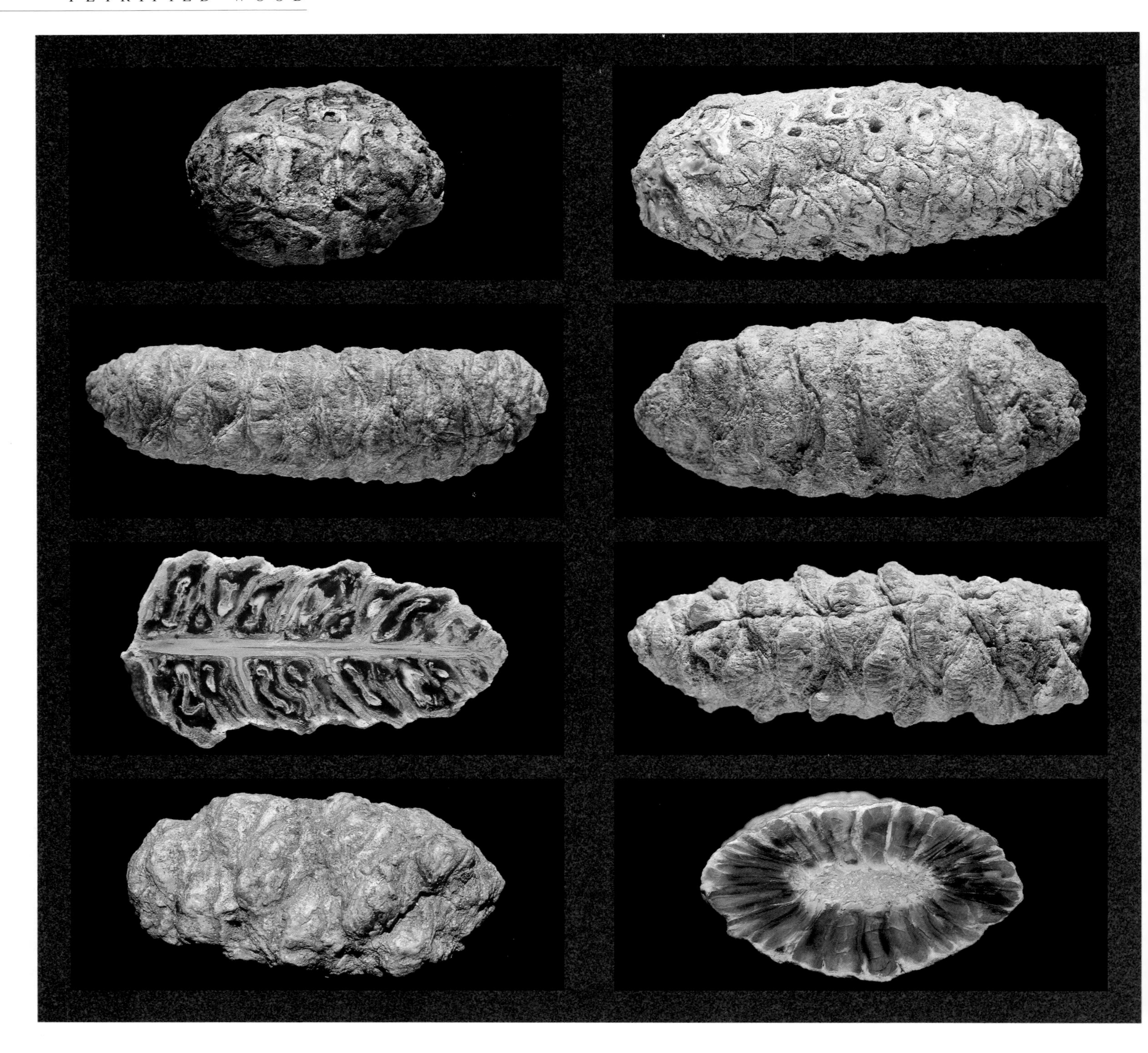

Idaho, USA [Bruneau]
Cenozoic; Tertiary
Pinus [pine] Daniels 6 cm

Patagonia, Argentina
Mesozoic; Jurassic
Pararaucaria Daniels 5 cm

Patagonia, Argentina
Mesozoic; Jurassic
Pararaucaria Jones 4 cm

Utah, USA [Henry Mountains]
Mesozoic; Jurassic fm: Morrison
Araucaria Hatch 3.5 cm

Colorado, USA [Grand Junction]
Mesozoic; Jurassic fm: Morrison
conifer anonymous 7 cm

Utah, USA [Henry Mountains]
Mesozoic; Jurassic fm: Morrison
Araucaria Hatch 4 cm

Patagonia, Argentina
Mesozoic; Jurassic
Pararaucaria Daniels 4.5 cm

Utah, USA [Henry Mountains]
Mesozoic; Jurassic fm: Morrison
Araucaria Hatch 2 cm

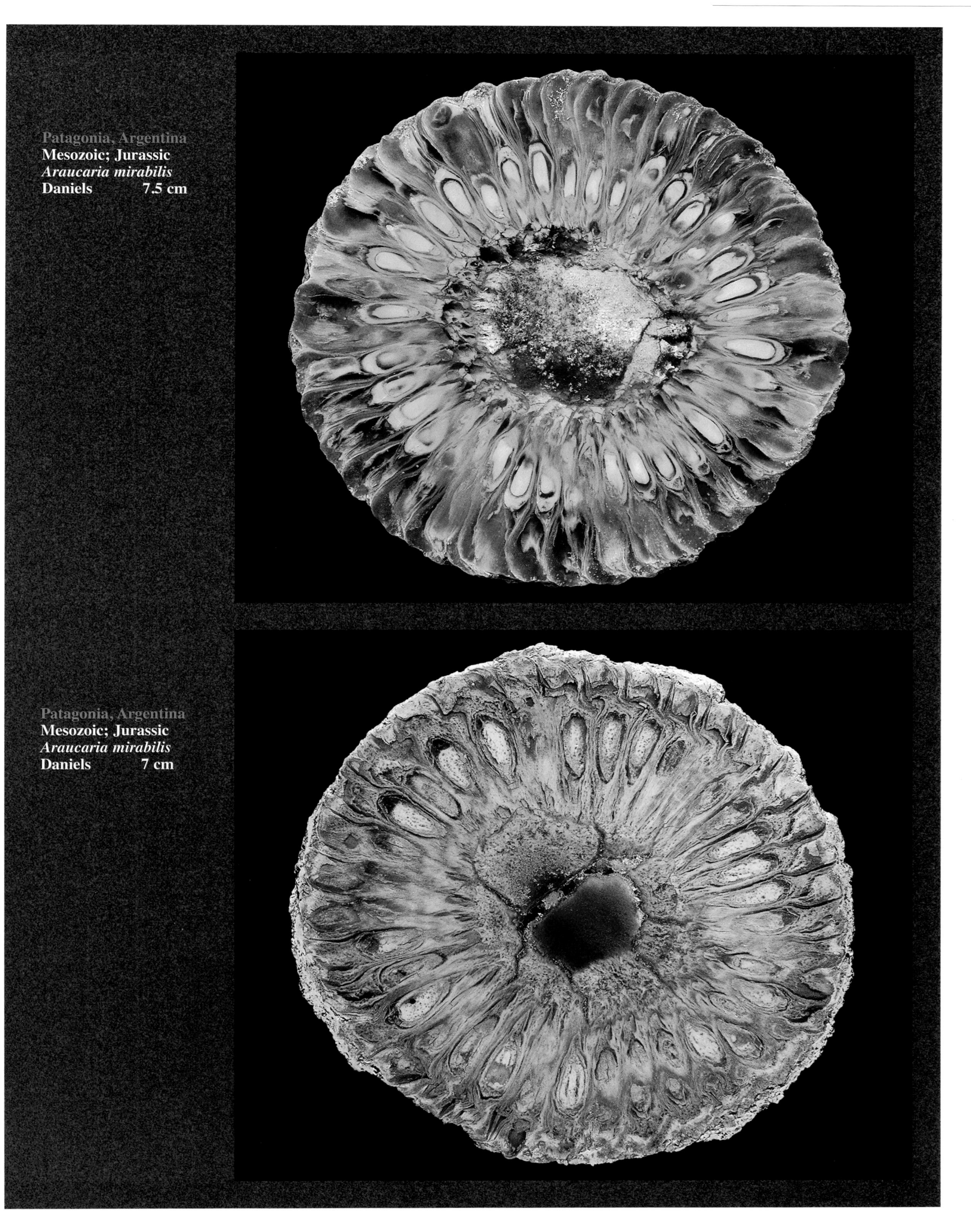

Patagonia, Argentina
Mesozoic; Jurassic
Araucaria mirabilis
Daniels 7.5 cm

Patagonia, Argentina
Mesozoic; Jurassic
Araucaria mirabilis
Daniels 7 cm

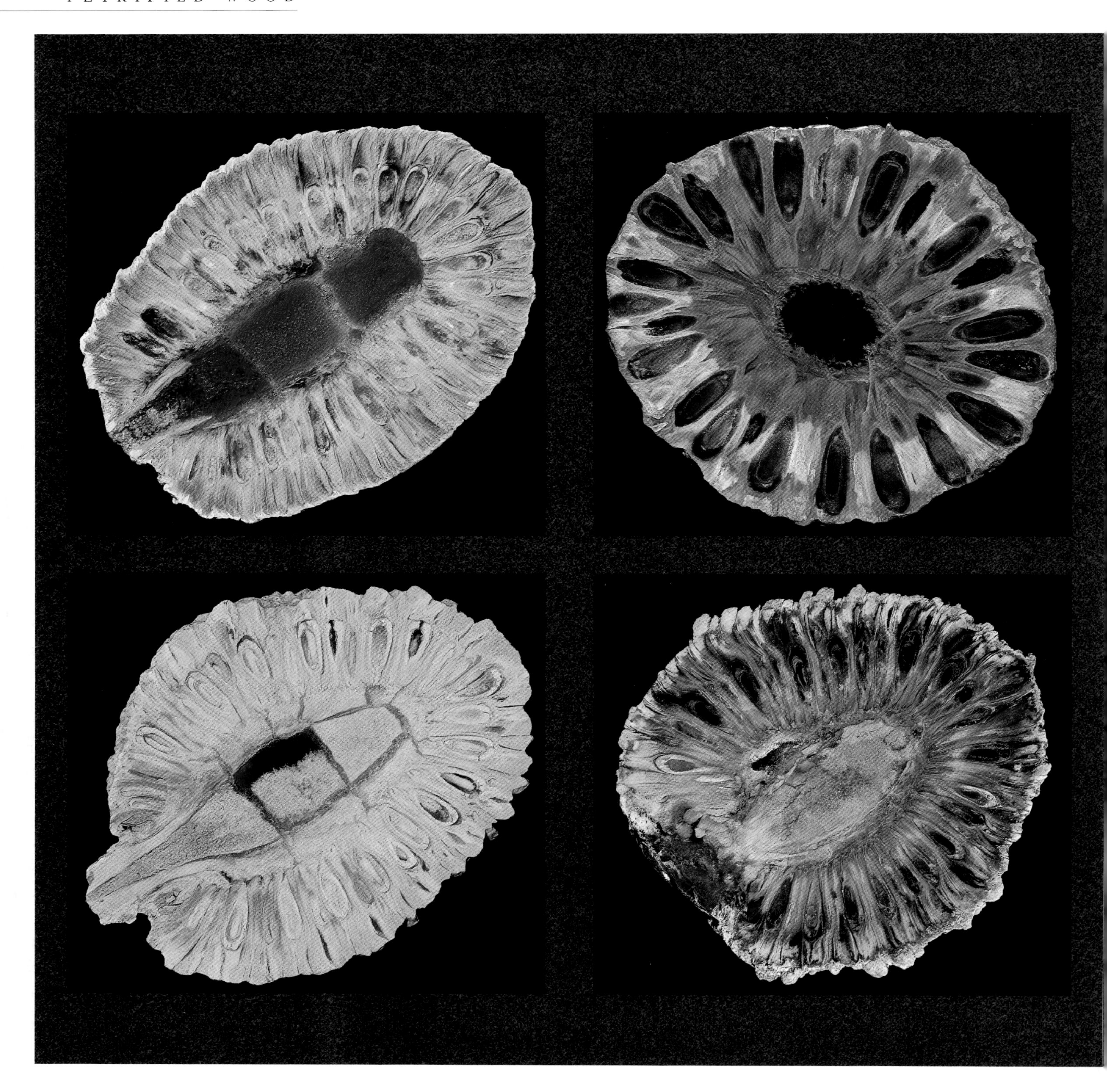

Patagonia, Argentina
Mesozoic; Jurassic
Araucaria mirabilis
Daniels 9 cm

Patagonia, Argentina
Mesozoic; Jurassic
Araucaria
Branson 5.5 cm

Patagonia, Argentina
Mesozoic; Jurassic
Araucaria mirabilis
Branson 8 cm

Patagonia, Argentina
Mesozoic; Jurassic
Araucaria mirabilis
Daniels 7.5 cm

Patagonia, Argentina
Mesozoic; Jurassic
Araucaria mirabilis
Daniels 6 cm

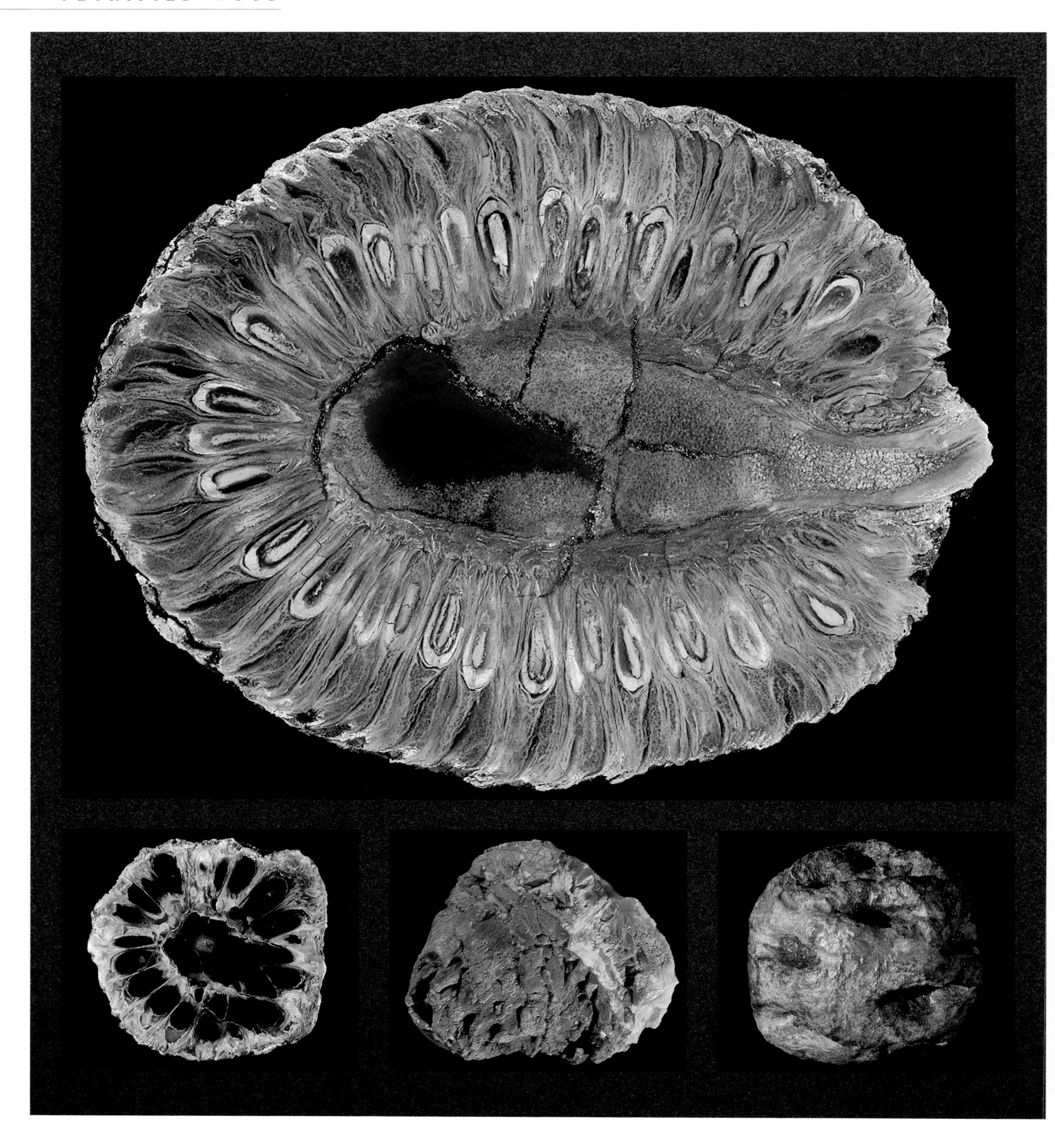

(TOP)
Patagonia, Argentina
Mesozoic, Jurassic
Araucaria mirabilis
Daniels 8.5 cm

(LOWER LEFT)
Wyoming, USA
Cenozoic; Tertiary; Eocene
conifer fm: Green River
Jones 5 cm

(LOWER MIDDLE)
Utah, USA [Henry Mountains]
Mesozoic; Jurassic
Cycadeoidea **fm: Morrison**
Hatch 3.5 cm

(LOWER RIGHT)
North Dakota, USA
Mesozoic; Cretaceous
***Sequoia* [redwood]**
Daniels 3.5 cm

Ferns, *Pentoxylon*, and *Hermanophyton* 7

"We have the receipt of fern-seed, we walk invisible."

William Shakespeare

1564-1616

King Henry IV

Ferns are an ancient plant group (Pteridophyta) which likely gave rise to the seed plants. Ferns were one of the dominant land plants during the Pennsylvanian and Mississippian periods. Fossil ferns also have been found in Devonian to modern strata. While ferns living today do not bear seeds, the fossil record contains evidence of "seed ferns" that bore seeds directly on the fronds (Pteridospermophyta).

Osmundaceae is a family with living members. They are unique among living ferns in that their rather small rhizomes or stems grow upright, rather than horizontally, supporting a cluster of fronds. The petrified leaf bases appear as "omegas" in polished cross-sections. Stems of most modern ferns grow underground and are generally weak and inconspicuous; roots arise sporadically. As is the case in general with ferns, the leaves or fronds of the Osmundaceae are large and pinnately compound and uncoil as they develop.

Tempskya is a form genus for the false trunks of a short to medium tree fern. The false trunk of *Tempskya* is composed of loosely organized leaf bases and stems encased in a dense mat of adventitious roots. A trunk looks similar to a poorly preserved palm trunk; the roots formed a conical false trunk that enabled the stem to assume an upright growth position. The roots were cannibalistic and attacked older leaf bases and roots. Known only from Lower Cretaceous rocks, a small "forest" of *Tempskya* with trunks preserved in an upright, growth position was discovered in Utah.

Pentoxylon is a gymnosperm of unknown affinity. The specimens shown are from the Jurassic Period of Australia. *Pentoxylon* was first identified from Jurassic formations in the Rajmahal Hills of India. It is characterized by a series of vascular segments arranged around central tissue. The number of vascular segments is typically five (hence the name), but can vary from plant to plant. They are somewhat similar to cycadeoids and apparently bore cones. The Australian fossils discussed here as "*Pentoxylon*" are so called because they appear closer in structure to described *Pentoxylon* than to any other known gymnosperm genus. It is anticipated that forthcoming analysis will result in new nomenclature for these fascinating and somewhat enigmatic fossils.

Hermanophyton is a stem genus of a poorly understood group of gymnosperms. When first described, *Hermanophyton* was thought to be a climbing vine. Recent finds, however, indicate it grew as a straight trunk. Branches are rare and split the stem into nearly equal halves. The outer surfaces of well-preserved specimens are covered with narrow-based leaf scars. No leaves assignable to the stem have been found. Paleobotanists await the discovery of leaves and additional stems to help unravel the relationship of *Hermanophyton* to other gymnosperms.

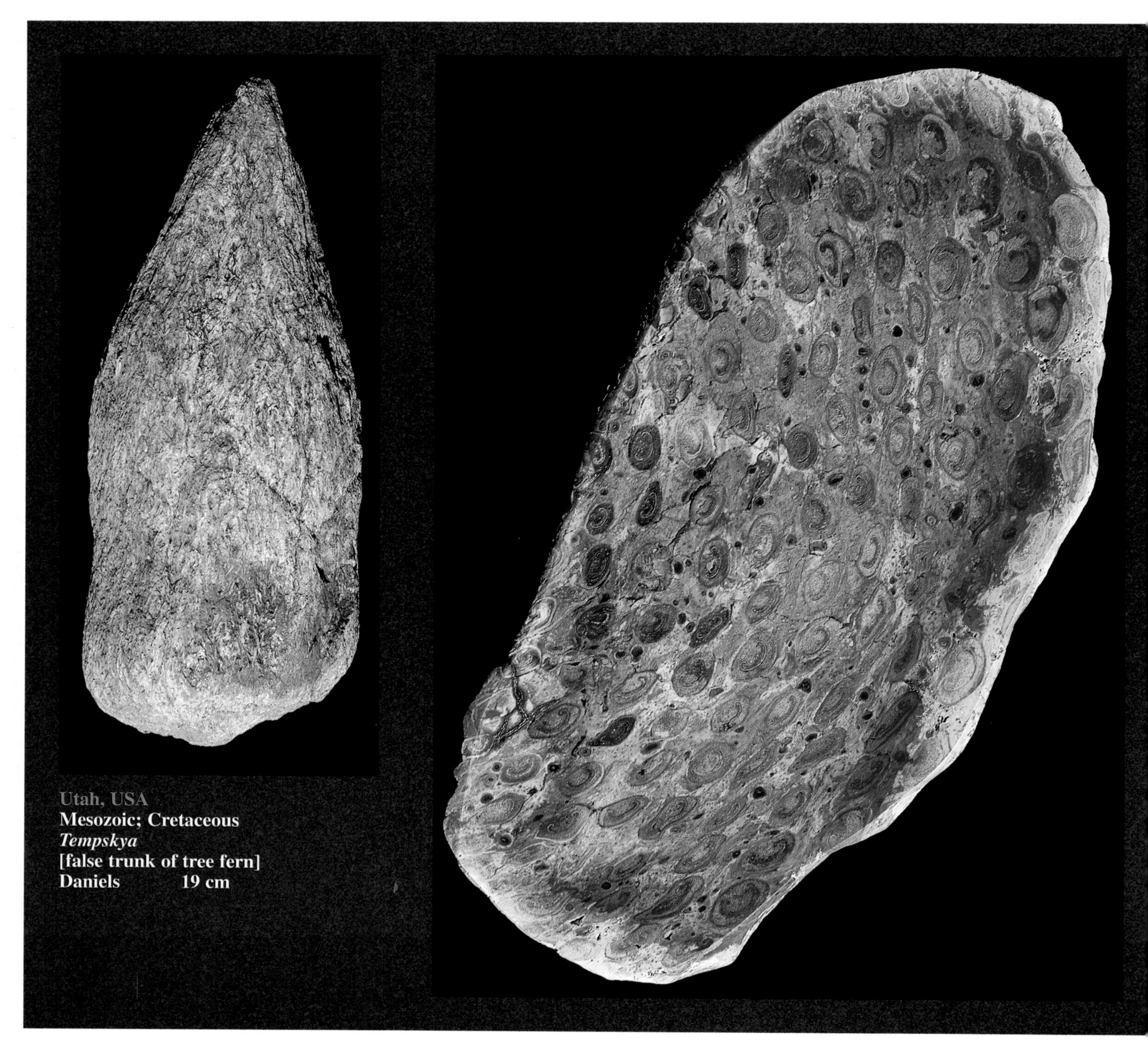

Utah, USA
Mesozoic; Cretaceous
Tempskya
[false trunk of tree fern]
Daniels 19 cm

Queensland, Australia
Mesozoic; Jurassic
Osmundacaulis
fm: Miles
Daniels 15.5 cm

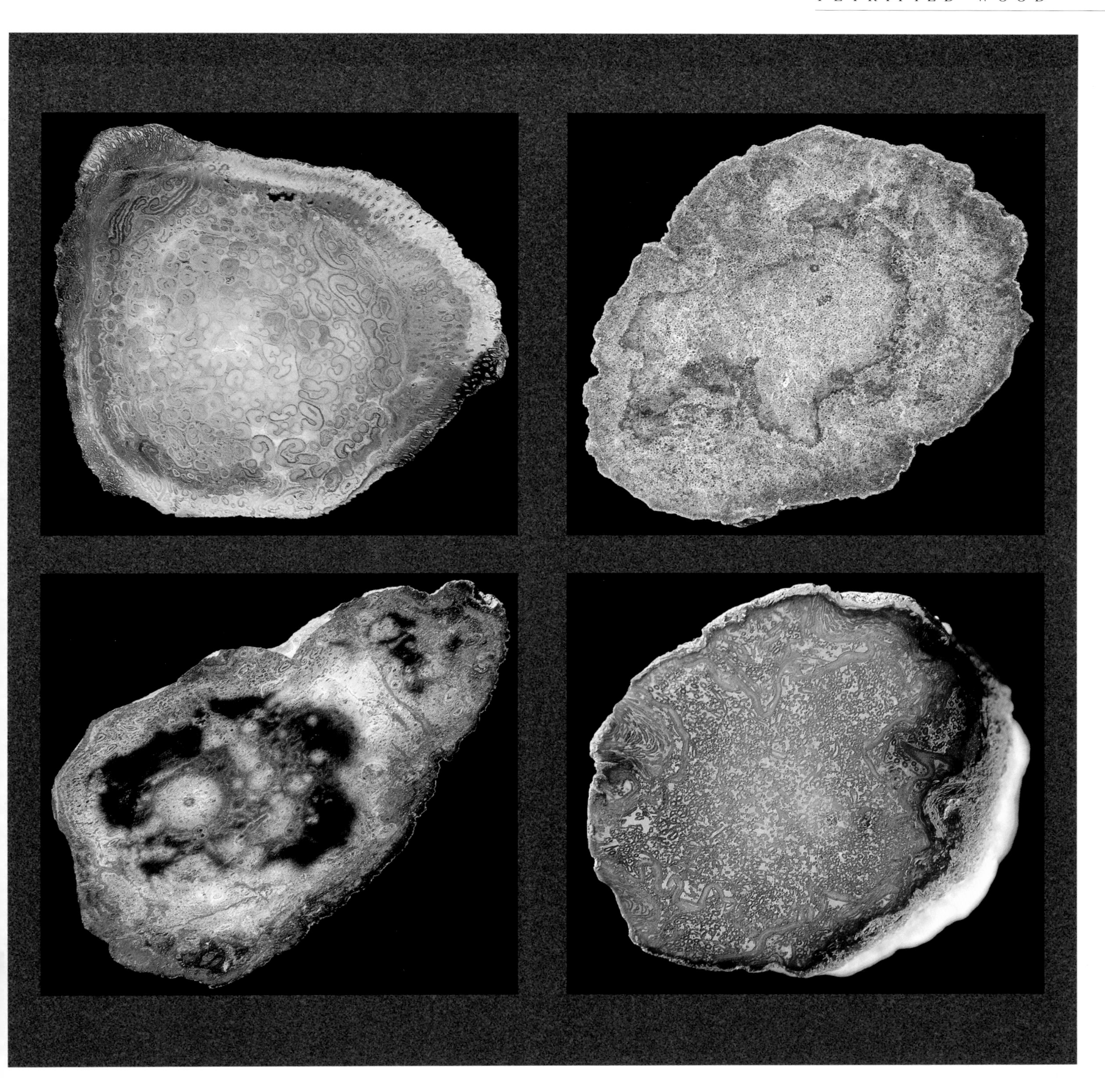

Maranhao, Brazil [Araguaina]
Paleozoic; Permian
Tietea singularis
Daniels 14.5 cm

Utah, USA
Mesozoic; Cretaceous
Tempskya
Daniels 18 cm

Colorado, USA
Mesozoic; Cretaceous
Tempskya
Daniels 18 cm

Texas, USA
Cenozoic; Tertiary; Eocene
Cyathodendron texanum
[tree fern]
Daniels 7 cm

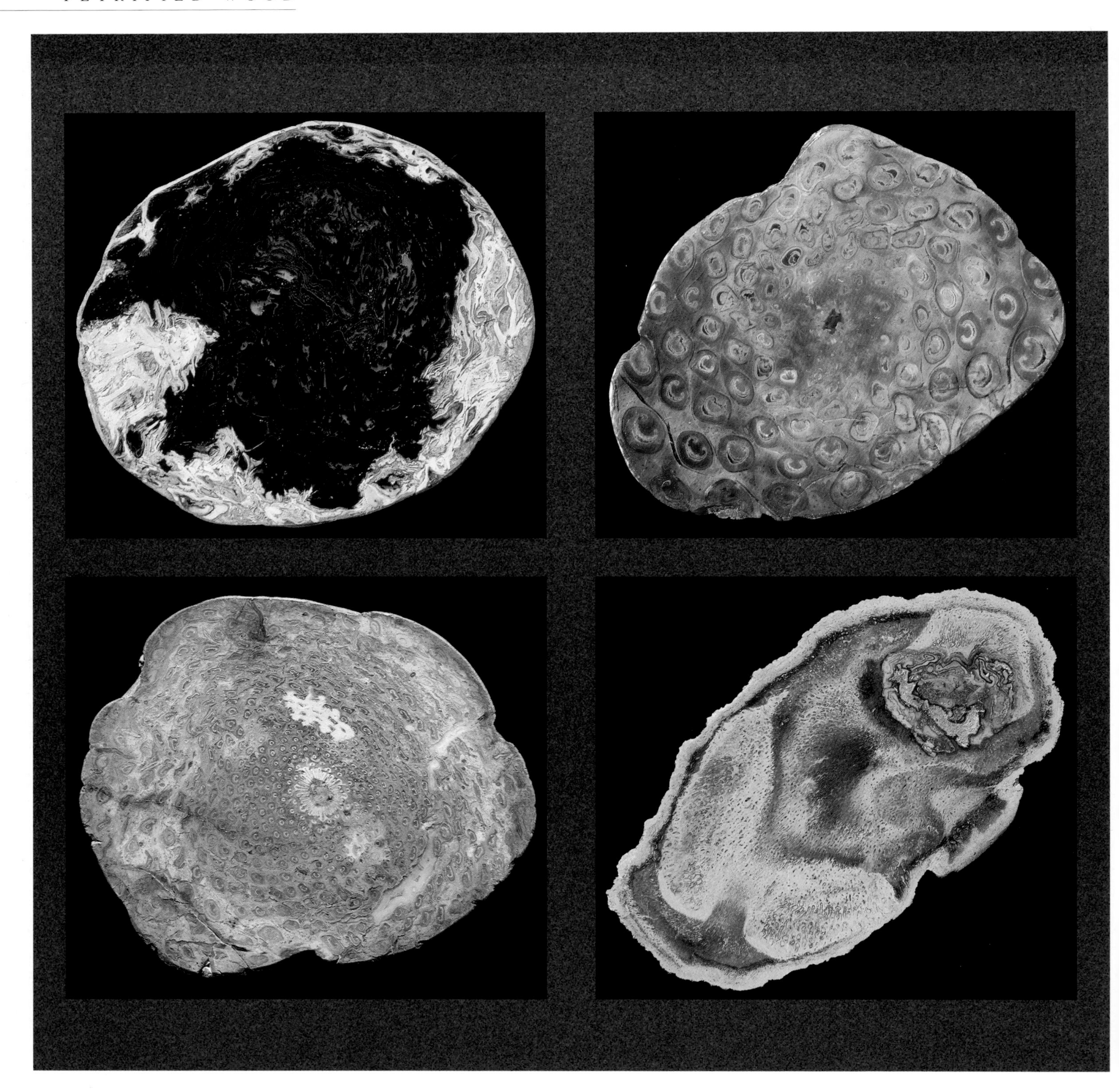

Queensland, Australia
Mesozoic; Jurassic
unidentified fern fm: Miles
Rigel 12.5 cm

Queensland, Australia
Mesozoic; Jurassic
unidentified fern fm: Miles
Rigel 7.5 cm

Queensland, Australia
Mesozoic; Jurassic
unidentified fern fm: Miles
Rigel 12 cm

Brazil
Paleozoic; Permian
tree fern
Cataldo 24.5 cm

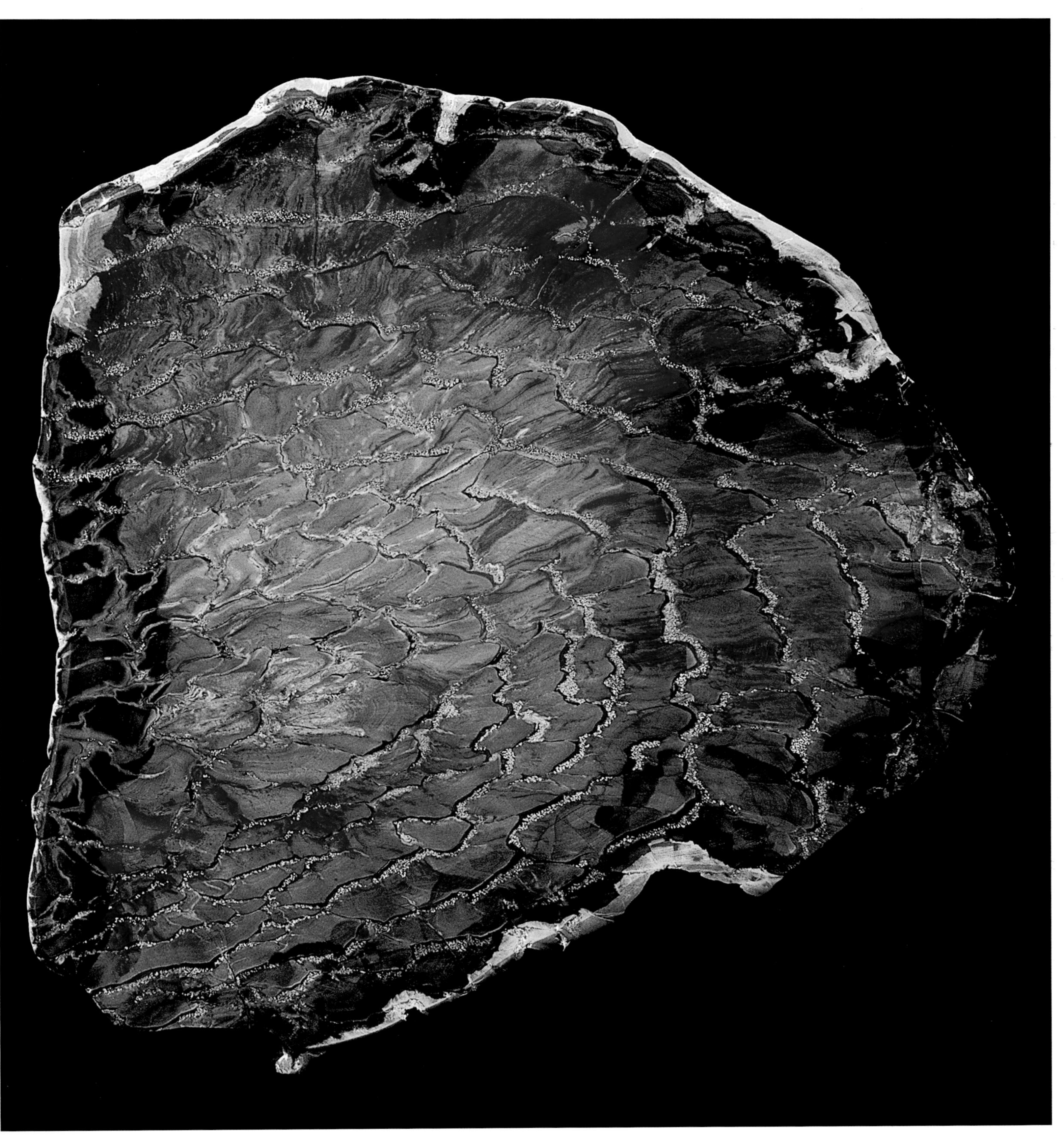

Queensland, Australia
Mesozoic; Jurassic
"Pentoxylon" **fm: Miles**
Daniels 23 cm

Queensland, Australia
Mesozoic; Jurassic
"Pentoxylon"
fm: Miles
Daniels 19 cm

Queensland, Australia
Mesozoic; Jurassic
"Pentoxylon"
fm: Miles
Rigel 25.5 cm

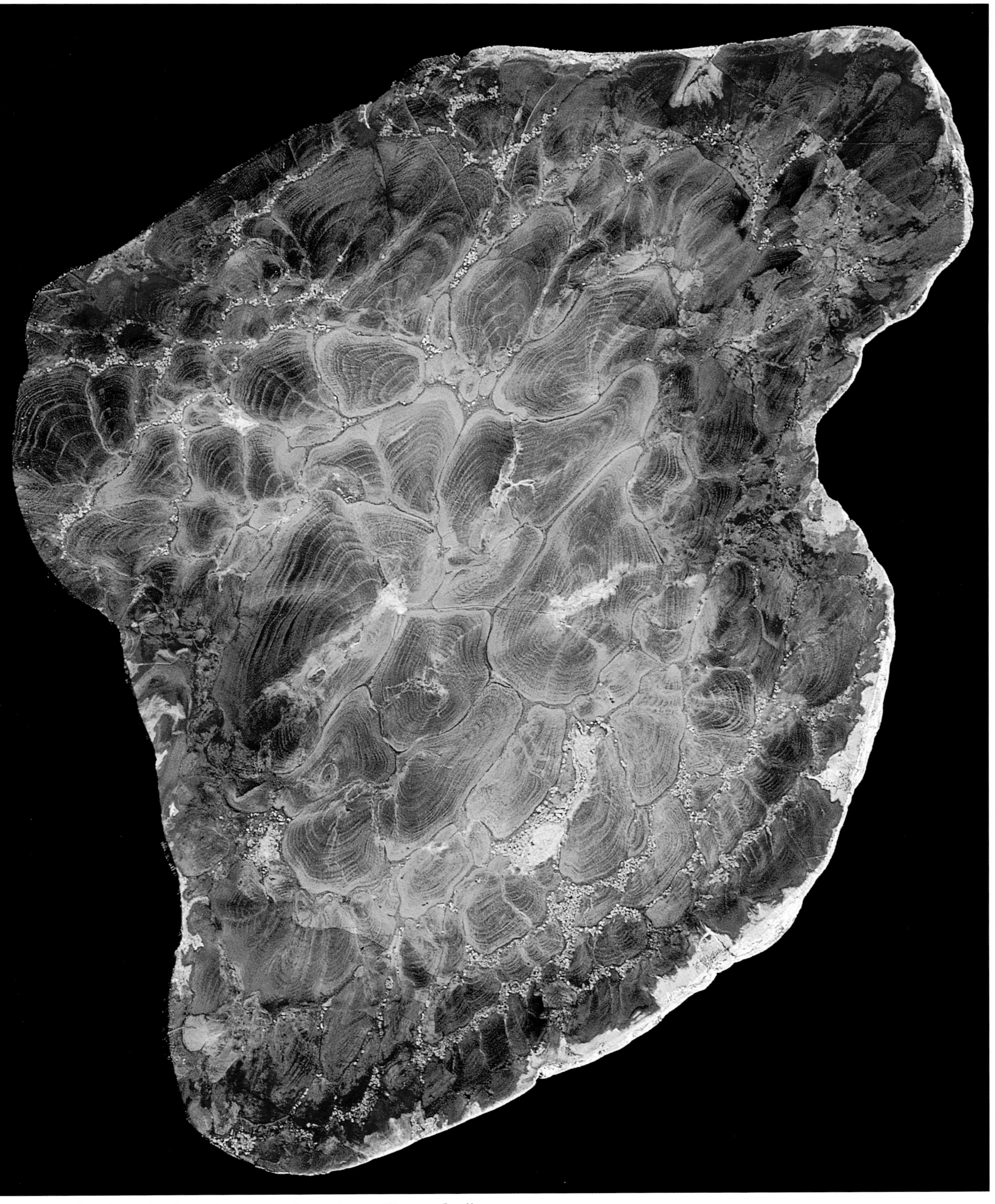

Queensland, Australia
fm: Miles

Mesozoic; Jurassic
Rigel

"Pentoxylon"
16.5 cm

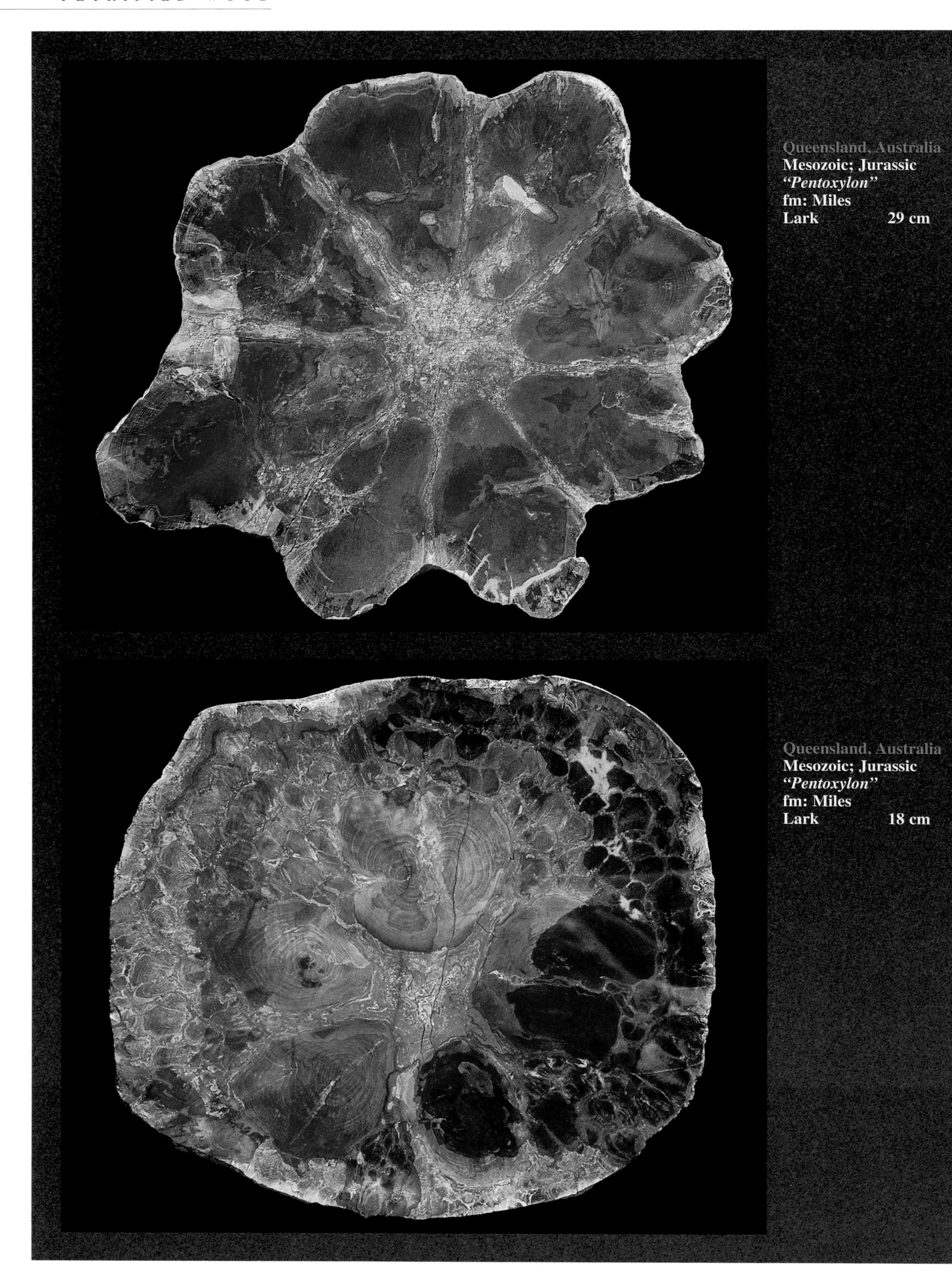

Queensland, Australia
Mesozoic; Jurassic
"Pentoxylon"
fm: Miles
Lark 29 cm

Queensland, Australia
Mesozoic; Jurassic
"Pentoxylon"
fm: Miles
Lark 18 cm

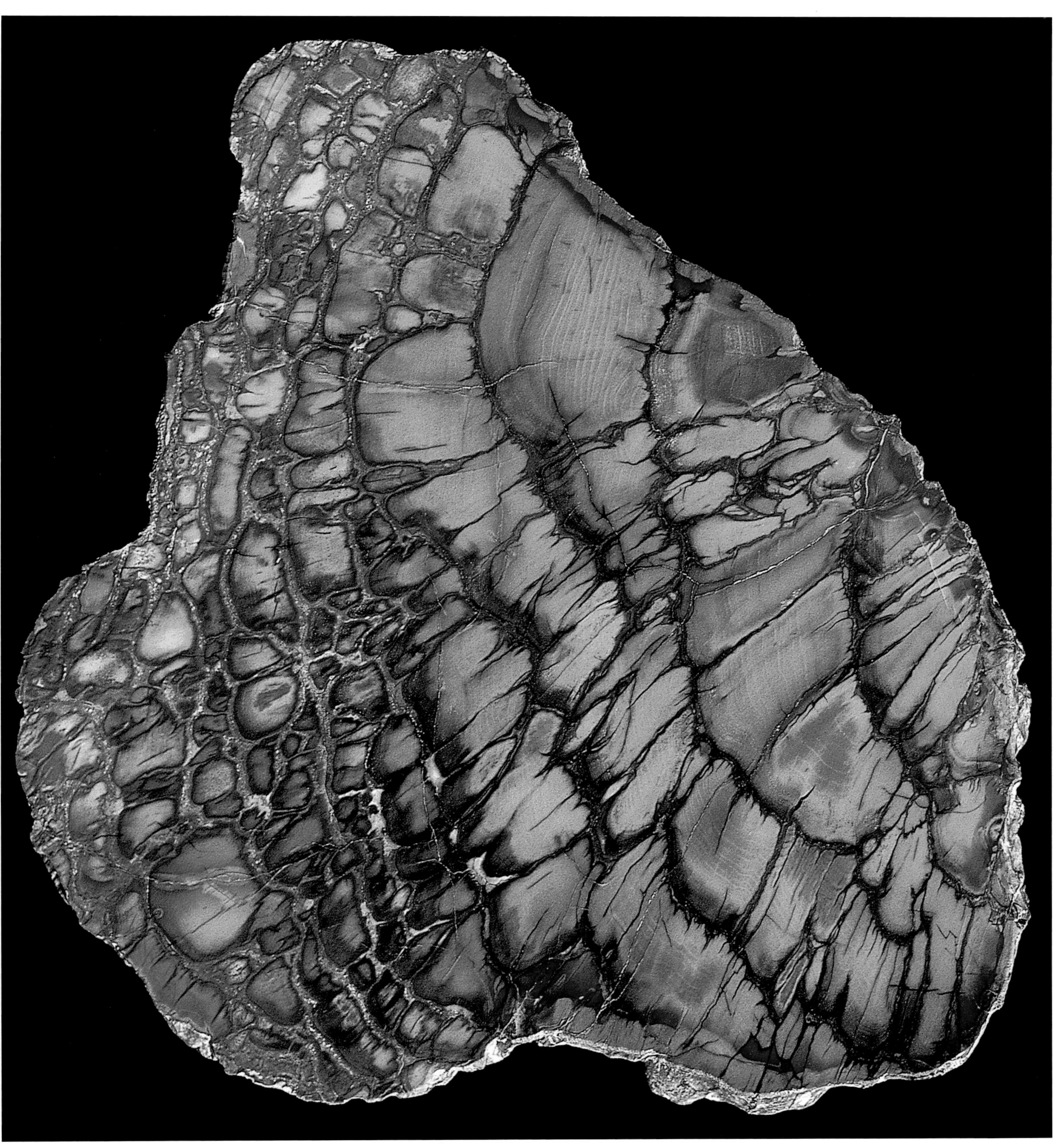

Queensland, Australia
Mesozoic; Jurassic
"Pentoxylon" **fm: Miles**
Rigel 28.5 cm

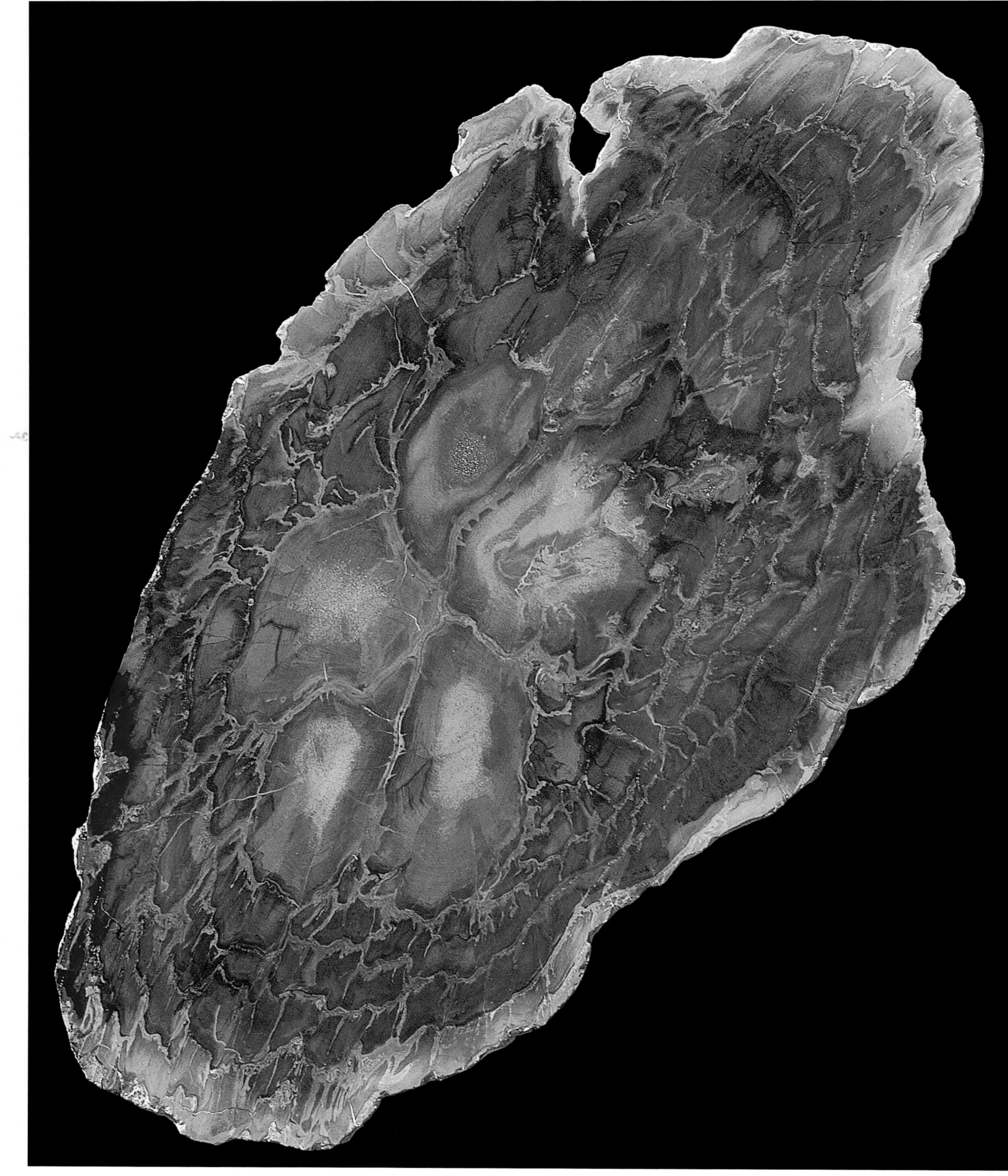

Queensland, Australia
fm: Miles

Mesozoic; Jurassic
Lark

"Pentoxylon"
38 cm

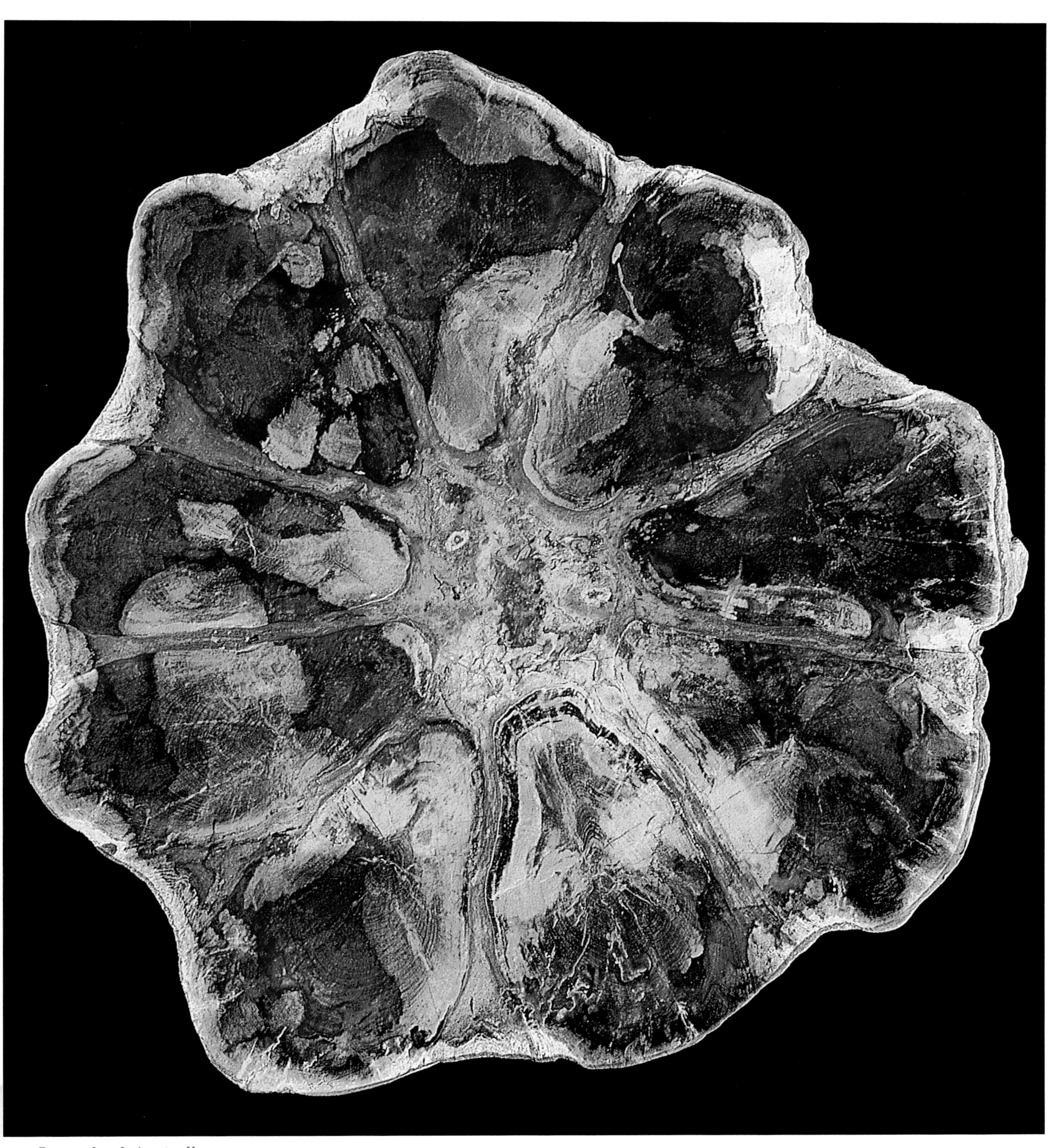

Queensland, Australia
Mesozoic; Jurassic
***"Pentoxylon"* fm: Miles**
Bennett 21 cm

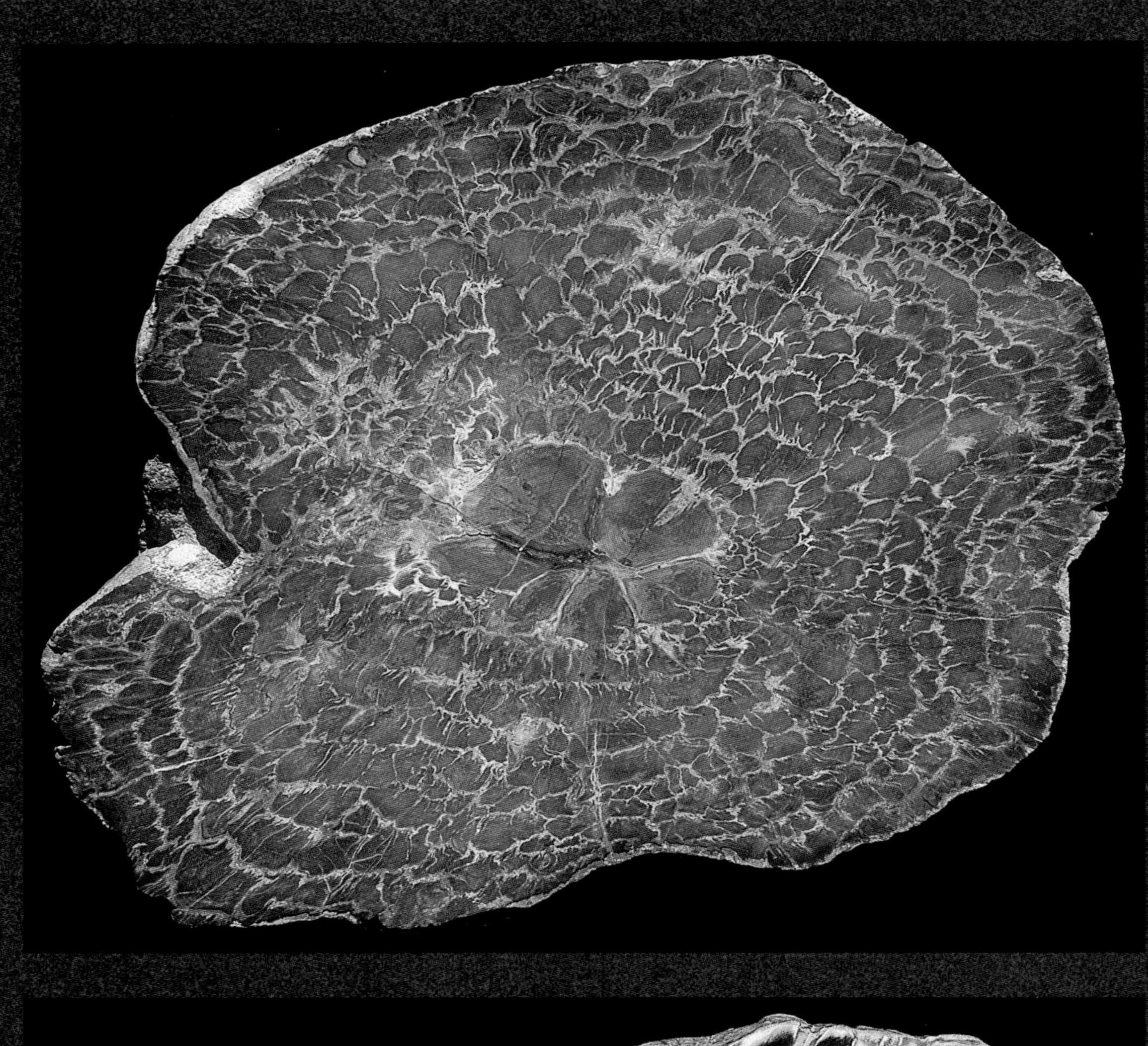

Queensland, Australia
Mesozoic; Jurassic
"Pentoxylon"
fm: Miles
Lark 28 cm

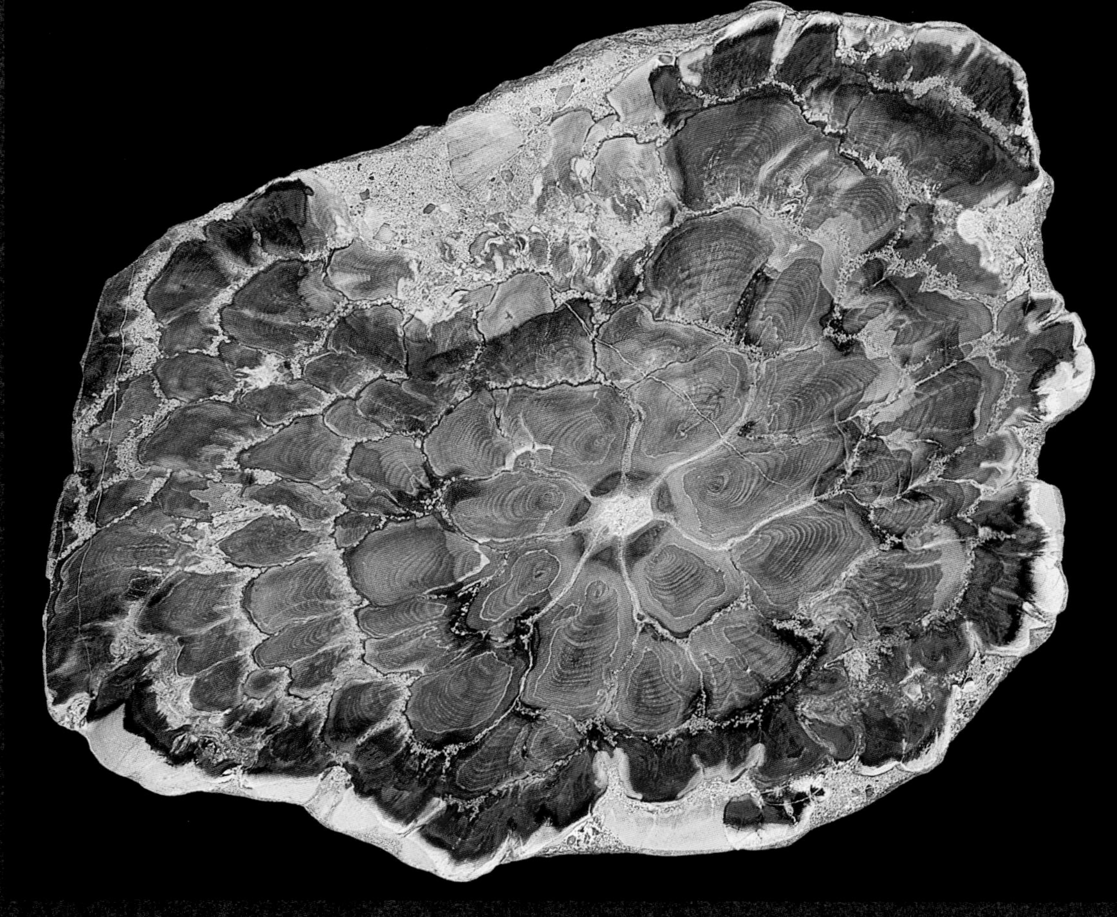

Queensland, Australia
Mesozoic; Jurassic
"Pentoxylon"
fm: Miles
Bennett 21 cm

Colorado, USA [Cortez]
Mesozoic; Jurassic
Hermanophyton **fm: Morrison**
Daniels 6 cm tall

Colorado, USA [Cortez]
Mesozoic; Jurassic
Hermanophyton **fm: Morrison**
Branson 6 cm

Colorado, USA [Cortez]
Mesozoic; Jurassic
Hermanophyton **fm: Morrison**
Sanchez 11 cm

Colorado, USA [Cortez]
Mesozoic; Jurassic
Hermanophyton **fm: Morrison**
Sanchez 10 cm

Colorado, USA [Cortez]
Mesozoic; Jurassic
Hermanophyton **fm: Morrison**
Sanchez 10 cm

Colorado, USA [Cortez]
Mesozoic; Jurassic
Hermanophyton **fm: Morrison**
Sanchez 8 cm

Colorado, USA [Cortez]
Mesozoic; Jurassic
Hermanophyton **fm: Morrison**
Daniels 9 cm

Colorado, USA [Cortez]
Mesozoic; Jurassic
Hermanophyton **fm: Morrison**
Branson 9 cm

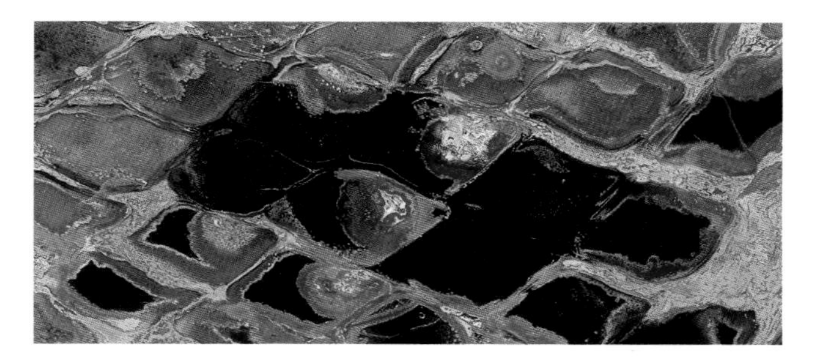

8

Cycads

"The unusual outlines of these rare fossils have probably never failed to arrest the attention of the lettered and the unlettered alike, from the remotest antiquity."
G.R. Wieland, 1906

The Jurassic Period is sometimes referred to as the Age of Cycads. Although cycads are grouped with the gymnosperms, they are not closely related to any other group of living plants. The plants have thick, stout trunks with an external armor of hardened, spirally arranged, leaf bases. Leaf bases form as fronds die off proceeding from lower to upper portions of the plant. The crown consists of large palm-like leaves. Some cycads have short, squat stems while others have taller, columnar stems. The name cycad comes from the Greek word "cyckos," which means palm-like. Cycads are not palms, although they are often mistaken as such. The Sego Palm, a common house plant, is actually a cycad and not a palm. The 185 living species of cycads are merely a relic of the large populations that once existed. At least one cycad, the South African *Encephalartos woodii*, is doomed to extinction since the few existing plants are all male. Cycads produce large cones. Male plants produce pollen cones; female plants produce seed cones. Some Central American cycads are pollinated by weevils hatched within male cones and attracted to female cones as they mature and release a pheromone. The weevils then unknowingly carry pollen from male to female plants, pollinating the ovules in the female cones. Other cycads are wind pollinated.

In this book, the term cycad is used to describe Cycadeoidales and Cycadales, the two orders of Division Cycadophyta. Cycadeoids belong to an order that may have evolved from seed ferns and has been extinct since the Cretaceous Period. Representatives of the Cycadales are found both as fossils and as living plants.

Cycads are of particular interest to collectors of petrified woods because of their rarity and the beauty of polished specimens of tangential sections cut through armor. These sections can reveal detailed diamond patterns in a multitude of colors. Cycads have been collected from Permian to modern formations. Petrified cycad trunks are noted particularly from the Jurassic and Cretaceous periods. Early collectors of cycad trunks often did not know what they were collecting, believing they had brought home a fossilized beehive, coral, mushroom, or mass of barnacles. Cycad trunks were seriously collected from localities on the Isle of Portland and the Isle of Wight in Great Britain in the late 19th century, and have been located in various places throughout Europe, North America, Central and South America, South Africa, India, and Australia. Unfortunately, collecting localities that once were noted for an abundance of cycad stems are now depleted. An important location historically was discovered by the scientific community in 1893 in the Black Hills of South Dakota. This Mesozoic rim of the Black Hills of South Dakota and Wyoming was noted by Professor Wieland as being "the most important of all the American cycad horizons." Within a few years, the entire area was depleted as all known specimens were removed. Many now reside in museum and university collections. Early in the century Yale University alone had over seven hundred specimens from the Black Hills, including those collected by Professor Wieland. A portion of the Black Hills area was set aside as a national monument to preserve the cycad fossils, but even there all exposed specimens are long gone. Fossil cycads in Utah were discovered later when uranium exploration reached into previously uncharted areas.

The pithy centers of most cycad varieties frequently rotted or compressed prior to petrification, allowing only the harder, leaf-base armor to survive. Because of the extreme rarity of fossilized cycads and their intense collection, complete trunks are difficult to acquire.

Plate from *American Fossil Cycads, Volume II*, 1916.

Patagonia, Argentina
Mesozoic; Jurassic/Cretaceous
cycad [transverse section]
Daniels 26 cm

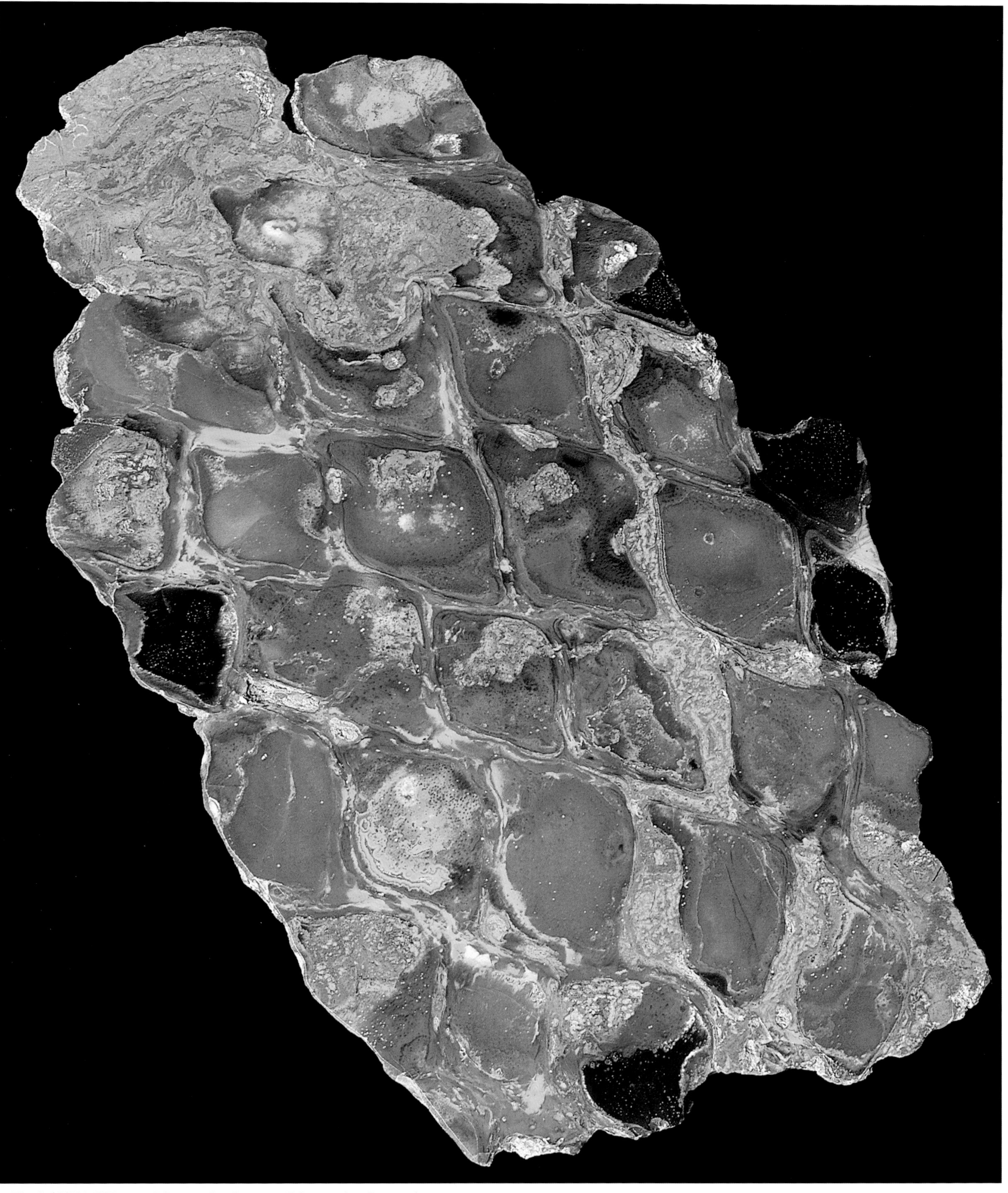

Utah, USA [Henry Mountains] Mesozoic; Jurassic
fm: Morrison Branson 24.5 cm
Cycadeoidea [tangential section through armor]

Patagonia, Argentina
Mesozoic; Jurassic/Cretaceous
Cycadeoidea **[tangential section through armor]**
Daniels 12.5 cm

Utah, USA [Henry Mountains]
Mesozoic; Jurassic fm: Morrison
Cycadeoidea **[tangential section through armor]**
Hatch 10 cm

Patagonia, Argentina
Mesozoic; Jurassic/Cretaceous
Cycadeoidea
Daniels 10.5 cm

Utah, USA [Henry Mountains]
Mesozoic; Jurassic fm: Morrison
Cycadeoidea **[tangential section through armor]**
Hatch 10 cm

Utah, USA [Henry Mountains]
Mesozoic; Jurassic fm: Morrison
Cycadeoidea **[tangential section through armor]**
Hatch 10 cm

Utah, USA [Henry Mountains]
Mesozoic; Jurassic fm: Morrison
Cycadeoidea **[tangential section through armor]**
Hatch 9 cm

Utah, USA [Henry Mountains]
Mesozoic; Jurassic
Cycadeoidea **fm: Morrison**
Daniels 53 pounds 42 cm

Utah, USA [Henry Mountains]
Mesozoic; Jurassic
Cycadeoidea **fm: Morrison**
Hatch 35 pounds 38 cm

Utah, USA [Henry Mountains]
Mesozoic; Jurassic
Cycadeoidea **fm: Morrison**
Hatch 57 pounds 35.5 cm

Patagonia, Argentina
Mesozoic; Jurassic/Cretaceous
Cycadeoidea
Daniels 9 cm

Utah, USA
[Henry Mountains]
Mesozoic; Jurassic
fm: Morrison
Cycadeoidea
[tangential section through armor]
Branson 24.5 cm

Patagonia, Argentina
Mesozoic; Jurassic/ Cretaceous
Cycadeoidea
[tangential section through armor]
Daniels 11.5 cm

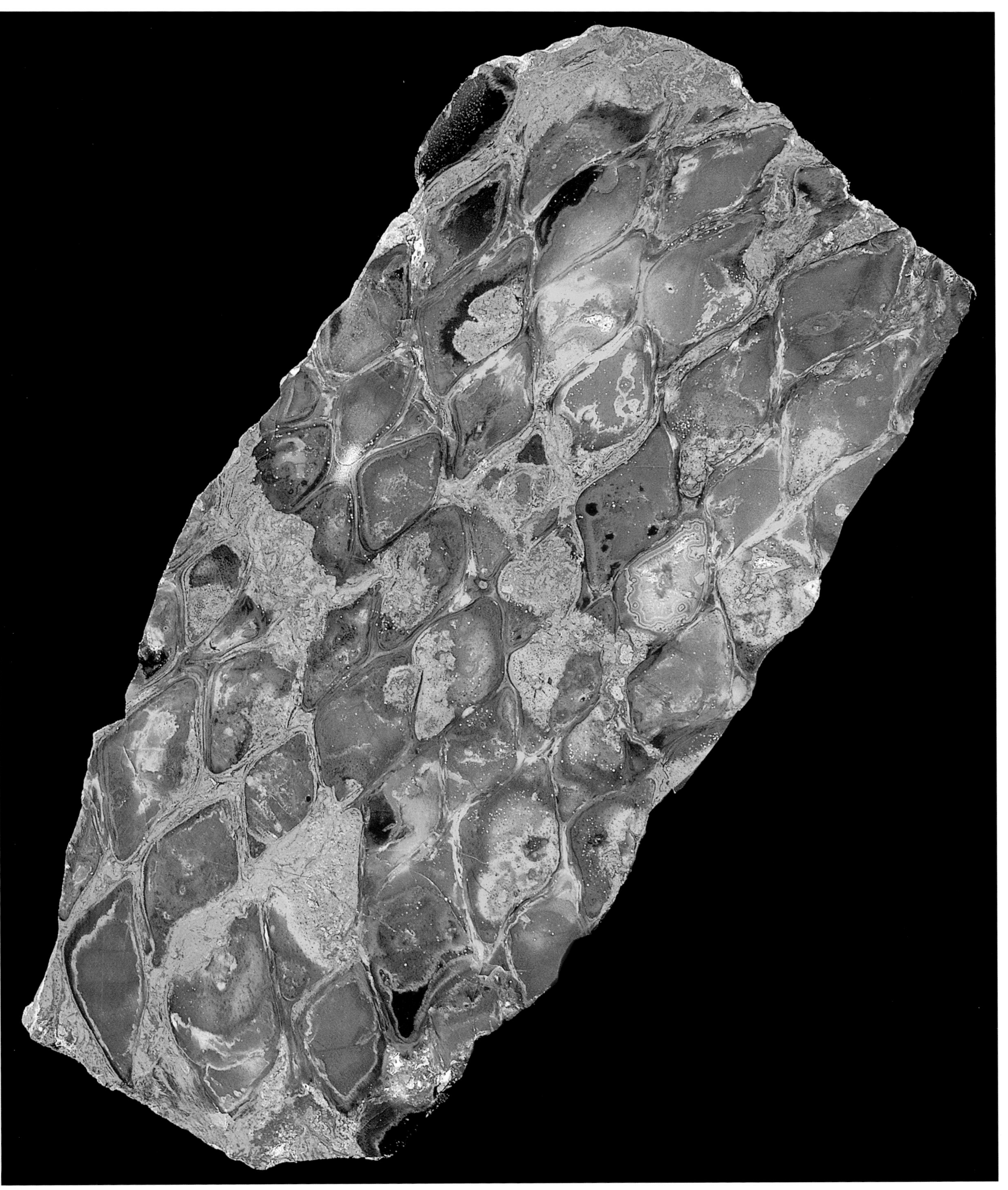

Utah, USA [Henry Mountains]
Mesozoic; Jurassic fm: Morrison
***Cycadeoidea* [tangential section through armor]**
Branson 31.5 cm

Patagonia, Argentina
Mesozoic; Jurassic/Cretaceous
Cycadeoidea
[tangential section through armor with cone]
Daniels 11 cm

Patagonia, Argentina
Mesozoic; Jurassic/Cretaceous
Cycadeoidea
[tangential section through armor with cone]
Daniels 15.5 cm

Patagonia, Argentina
Mesozoic; Jurassic/Cretaceous
Cycadeoidea **[tangential section through armor with cone]**
Daniels 15.5 cm

Patagonia, Argentina
Mesozoic; Jurassic/Cretaceous
Cycadeoidea **[tangential section through armor]**
Daniels 13.5 cm

Appendix

Following are some of the statutes of the United States dealing with the collection of petrified wood on national lands. These are not intended to be a complete guide; they are offered for informational purposes only. None of the rules and regulations promulgated in furtherance of these statutes have been included.

Laws found in Title 43 pertain to the Department of Interior which is responsible for lands administered by the Bureau of Land Management and the Bureau of Reclamation.

43 C.F.R. Section 3622.1 General

(a) Persons may collect limited quantities of petrified wood for noncommercial purposes under terms and conditions consistent with the preservation of significant deposits as a public recreational resource.

(b) The purchase of petrified wood for commercial purposes is provided for in Sec. 3610.1 of this title.

43 C.F.R. Section 3622.3 Designation of Areas.

(a) All public lands administered by the Bureau of Land Management and the Bureau of Reclamation are open to or available for free use removal of petrified wood unless otherwise provided for by notice in the Federal Register. Free use areas under the jurisdiction of said Bureaus may be modified or canceled by notices published in the Federal Register.

(b) The heads of other Bureaus in the Department of the Interior may publish in the Federal Register designations, modifications or cancellations of free use areas for petrified wood on lands under their jurisdiction.

(c) The Secretary of the Interior may designate, modify or cancel free use areas for petrified wood on public lands which are under the jurisdiction of other Federal departments or agencies, other than the Department of Agriculture, with the consent of the head of other Federal departments or agencies concerned, upon publication of notice in the Federal Register.

43 C.F.R. Section 3622.4 Collection Rules

(a) General. The authorized officer shall control the removal without charge of petrified wood from public lands using the following criteria:

(1) The maximum quantity of petrified wood that any one person is allowed to remove without charge per day is 25 pounds in weight plus one piece, provided that the maximum total amount that one person may remove in one calendar year shall not exceed 250 pounds. Pooling of quotas to obtain pieces larger than 250 pounds is not allowed.

(2) Except for holders of permits issued under subpart 3621 of this title to remove museum pieces, no person shall use explosives, power equipment, including, but not limited to, tractors, bulldozers, plows, power-shovels, semi-trailers or other heavy equipment for the excavation or removal of petrified wood.

(3) Petrified wood obtained under this section shall be for personal use and shall not be sold or bartered to commercial dealers.

(4) The collection of petrified wood shall be accomplished in a manner that prevents unnecessary and undue degradation of lands.

(b) Additional rules. The head of the agency having jurisdiction over a free use area may establish and publish additional rules for collecting petrified wood for noncommercial purposes to supplement those included in Sec. 3622.4(a) of this title.

Title 36 of the Code of Federal Regulations codifies the laws enacted by the United States Congress pertaining to lands under control of the Department of Agriculture, which includes all national forests.

36 C.F.R. Section 228.62 Free Use

(a) Application. An application for a free-use permit must be made with the appropriate District Ranger's office.

(b) Term. Permits may be issued for periods not to exceed 1 year and will terminate on the expiration date unless extended by the authorized officer as in Sec. 228.53(b). However, the authorized officer may issue permits to any local, State, Federal, or Territorial agency, unit or subdivision, including municipalities and county road districts, for periods up to 10 years.

(c) Removal by agent. A free-use permittee may extract the mineral materials through a designated agent provided that the conditions of the permit are not violated. No part of the material may be used as payment for the services of an agent in obtaining or processing the material. A permit may be issued in the name of a designated agent for those entities listed in Sec. 228.62(d)(1), at the discretion of the authorized officer, provided there is binding agreement in which the entity retains responsibility for ensuring compliance with the conditions of the permit.

(d) Conditions. Free-use permits may be issued for mineral materials to settlers, miners, residents, and prospectors for uses other than commercial purposes, resale, or barter (16 U.S.C. 477). Free-use permits may be issued to local, State, Federal, or Territorial agencies, units, or subdivisions, including municipalities, or any association or corporation not organized for municipalities, or any association or corporation not organized for profit, for other than commercial or industrial purposes or resale (30 U.S.C. 601). Free-use permits may not be issued when, in the judgment of the authorized officer, the applicant owns or controls an adequate supply of mineral material in the area of demand. The free-use permit, issued on a Forest Service-approved form, must include the basis for the free-use as well as the provisions governing the selection, removal, and use of the mineral materials. No mineral material may be removed until the permit is issued. The permittee must notify the authorized officer upon completion of mineral material removal. The permittee must complete the reclamation prescribed in the operating plan (Sec. 228.56).

(1) A free-use permit may be issued to any local, State, Federal, or Territorial agency, unit, or subdivision, including municipalities and county road districts, without limitations on the number of permits or on the value of the mineral materials to be extracted or removed.

(2) A free-use permit issued to a nonprofit association, corporation, or individual may not provide for the removal of mineral materials having a volume exceeding 5,000 cubic yards (or weight equivalent) during any period of 12 consecutive months.

(e) Petrified wood. A free-use permit may be issued to amateur collectors and scientists to take limited quantities of petrified wood for personal use. The material taken may not be bartered or sold. Free-use areas may be designated within which a permit may not be required. Removal of material from such areas must be in accord with rules issued by the authorized officer and posted on the area. Such rules must also be posted in the District Ranger's and Forest Supervisor's offices and be available upon request. The rules may vary by area depending on the quantity, quality, and accessibility of the material and the demand for it.

Bibliography

American Fossil Cycads
G.R. Wieland
Carnegie Institution of Washington
Washington, D.C. 1906

American Fossil Cycads, Volume II: Taxonomy
G.R. Wieland
Carnegie Institution of Washington
Washington, D.C. 1916

Araucaria
Ulrich Dernbach
D'Oro-Verlag
Germany 1992

Banded Agates: Origins and Inclusions
Roger K. Pabian and Andrejs Zarins
University of Nebraska
Lincoln, Nebraska 1994

Bennettitales of the English Wealden
Joan Watson and Caroline A. Sincock
The Palaeontographical Society
London, England 1992

The Biology and Evolution of Fossil Plants
T. N. Taylor and E. L. Taylor
Prentice-Hall
Englewood Cliffs, New Jersey 1993

The Book of Agates
Leland Quick
Chilton Book Company
Radnor, Pennsylvania 1963

Canyonlands Country: A Guidebook of the Four Corners Geological Society
James E. Fassett, Editor
Four Corners Geological Society
Durango, Colorado 1975

The Cerro Cuadrado Petrified Forest
G.R. Wieland
Carnegie Institution of Washington
Washington, D.C. 1935

Common Fossil Plants of Western North America
William D. Tidwell
Brigham Young University Press
Provo, Utah 1975
2nd Edition
Smithsonian Institution Press
Washington, D.C. 1998

Dana's Manual of Mineralogy
James D. Dana, as revised by Cornelius S. Hurlbut, Jr.
John Wiley & Sons, Inc.
New York, New York 1959

Dana's New Mineralogy: The System of Mineralogy of James Dwight Dana and Edward Salisbury Dana
Richard V. Gaines
John Wiley & Sons
New York, New York 1997

The Darwin Reader
Marston Bates and Philip S. Humphrey, Editors
Charles Scribner's Sons
New York, New York 1956

Dawn of the Dinosaurs: The Triassic in Petrified Forest
Robert A. Long and Rose Houk
Petrified Forest Museum Association
Petrified Forest, Arizona 1988

The Enigma of Angiosperm Origins
Norman F. Hughes
Cambridge University Press 1994

Fire, Where Does it Come From?
Robert C. Barnes
The International Opal Journal, Volume 1, Number 1
Reno, Nevada 1977

Flowering Plant Origin, Evolution & Phylogeny
David Winship Taylor, editor
Chapman & Hall 1996

Gemstones of North America, Volume 3
John Sinkankas
Geoscience Press
Tucson, Arizona 1997

Geological Factors and the Evolution of Plants
Bruce Tiffney, editor
Yale University Press 1985

Geologic History of Utah
Lehi F. Hintze
Brigham Young University
Provo, Utah 1988

Identification of Modern and Tertiary Woods
A. C. Barefoot
Clarendon Press 1982

Identifying Wood: Accurate Results With Simple Tools
R. Bruce Hoadley
The Taunton Press
Newtown, Connecticut 1990

Integrated Principles of Zoology
Cleveland P. Hickman, *et. al.*
Times Mirror/Mosby College Publishing
St. Louis, Missouri 1988

Manual of Mineralogy: After James D. Dana
Cornelius Klein and Cornelius S. Hurlbut, Jr.
John Wiley & Sons
New York, New York 1993

Mineralogy for Amateurs
John Sinkankas
Van Nostrand Reinhold Company
New York, New York 1964

Minerals: An Illustrated Exploration of the Dynamic Worlds of Minerals and Their Properties
George W. Robinson, Ph.D.
Simon & Schuster
New York, New York 1994

On The Upper Jurassic Stem: Hermanophyton And Its Species From Colorado and Utah, USA
William D. Tidwell and Sidney R. Ash
Palaeontographica
Stuttgart, Germany 1990

Opals
P.J. Darragh, A.J. Gaskin, and J.C. Sanders
Scientific American, Volume 24 1976

Paleobotany and the Evolution of Plants
Wilson N. Stewart and Gar W. Rothwell
Cambridge University Press
New York, New York 1993

Paleontology and Geology of the Dinosaur Triangle
Walter R. Averett, Editor
Museum of Western Colorado
Grand Junction, Colorado 1987

Petrified Forests of Yellowstone
Erling Dorf
Division of Publications, National Park Service
Washington, D.C.

Petrified Forest: The Story Behind the Scenery
Sidney Ash
Petrified Forest Museum Association
Petrified Forest, Arizona 1985

Petrified Forests: The World's 31 Most Beautiful Petrified Forests
Ulrich Dernbach
D'Oro-Verlag
Germany 1996

Plant Form and Function
Gerard J. Tortora, *et. al.*
The Macmillan Company
Toronto, Canada 1970

Protoyucca shadishii gen. et sp. nov., An Arborescent Monocotyledon with Secondary Growth from the Middle Miocene of Northwestern Nevada, U.S.A.
William D. Tidwell and Lee R. Parker
Review of Palaeobotany and Palynology, 62 (1990)
Elsevier Science Publishers B.V., Amsterdam

The Red Notebook of Charles Darwin
Sandra Herbert, Editor
Cornell University Press
Ithaca, New York 1980

Roadside Geology of Oregon
David Alt and Donald Hyndman
Mountain Press Publishing Company
Missoula, Montana 1978

Roadside Geology of Utah
Halka Chronic
Mountain Press Publishing Company
Missoula, Montana 1990

The System of Mineralogy of James Dwight Dana and Edward Salisbury Dana Volume III: Silica Minerals
Clifford Frondel
John Wiley and Sons, Inc. New York 1962

Glossary

agate: chalcedony with colored bands or fortifications

agatize: turn to agate

amber: fossilized tree resin

angiosperm: flowering plant with seeds which form within an ovary; the dominant plant of the modern era

brecciated: a material, such as chalcedonized, silicified, or jasperized wood, which has been broken into irregular fragments and then resealed by silica solutions; found often in petrified wood from the Yellow Cat, Utah, area

calcite: the common mineral form of calcium carbonate, a component of many fossils

carnelian: a red, orange-red, or brownish-red form of chalcedony

carnelian wood: petrified wood consisting largely of carnelian

cast: formed when a cavity left by a decayed piece of organic material is filled by another material; this other material can be mud or sand or can be largely composed of silicas, which can result in a cast of chalcedony or agate; in a true cast there will be no evidence of cell structure or organic materials; often a cast is a partial cast—part chalcedony with remnants of permineralized material, either intact or brecciated and suspended within

chalcedonize: turn to chalcedony

chalcedony: silicon dioxide, often without patterns or inclusions; a form of cryptocrystalline quartz

conglomerate: rock composed of rounded, water-worn fragments of older rock with sand in between

copal: fossilized tree resin, younger than amber and not as stable

cotyledon: the earliest leaf arising from a seed

cross or transverse section: a cut made perpendicular to the long axis of a stem, branch, or trunk

cycad: a gymnospermous, cone-bearing plant consisting of a thick, stout trunk (some are short and squat while others have taller, columnar stems) often composed of an armor of hardened, spirally arranged, leaf bases that form as fronds proceeding from the lower to upper portions of the plant die off; the crown consists of large, palm-like leaves

Cycadales: an order of Cycadophyta with both living and extinct, fossil representatives; cones occur at the apex of the stem

Cycadeoidales: an order of Cycadophyta with only extinct, fossil representatives; cones are embedded among the leaf bases

Cycadophyta: a division of gymnosperms which includes the orders Cycadales and Cycadeoidales; cycads

dicotyledon: an angiosperm in which the seed germinates to two leaves and the leaves have complex, net-like venation; abbreviated as dicot

double-hearted, triple-hearted, and so on: a petrified wood specimen revealing a multiple set of annual rings, representing a cross-section of a fork in a branch or trunk

formation: a sedimentary bed or series of beds sufficiently homogeneous or distinctive to be mapped at a scale of 1:25,000

genera: plural of genus

genus: a classification of plants or animals with common distinguishing characteristics

gymnosperm: a seed plant with seeds borne naked in the cone; includes seed ferns, cycads, ginkgos, and conifers

gypsum: hydrous calcium sulfate, a common mineral

hardwood: deciduous angiospermous wood with compact texture

inflorescence: a cluster of flowers or reproductive organs

jasper: an opaque cryptocrystalline quartz, SiO_2

jasperize: turn into jasper

leaf scar: a scar left where the leaf had been attached

limestone: a sedimentary rock consisting predominantly of calcium carbonate

macerate: to soften and wear away

member: subdivision of a geologic formation

monocotyledon: an angiosperm in which the seed germinates to one leaf (as in grasses, lilies, and palms) and the leaves have parallel veins; abbreviated as monocot

mudstone: a sedimentary rock formed from mud and composed of particles smaller than silt

opal: $SiO_2 \bullet n\ H_2O$; a form of silicon dioxide; hydrated silica, softer than quartz, less tough than chalcedony

opalize: turn to opal

permineralize: the process whereby soluble minerals enter the cells and fill the spaces between cells, resulting in a hardened mineral which contains and supports original tissues

petrify: to permineralize or to replace the normal cells of organic matter with silica or other mineral deposit

pinnate: with similar parts arranged on opposite sides of an axis

quartz: a crystalline form of silica; SiO_2 (silicon dioxide)

red bed: red sedimentary layer usually of siltstone or mudstone

radial section: a cut made lengthwise through a stem, branch, or trunk extending from the center to the periphery

round: a term for a specimen of petrified wood which includes the near complete branch or trunk circumference; "full round" will be very nearly 360° complete

sandstone: sedimentary rock formed predominantly from sand with particles larger than silt

shipworm wood: petrified wood which contains silicified bore tunnels of various bivalve mollusks such as *Teredo* and *Bankia*, commonly known as shipworms

silex: silica

silica: silicon dioxide (SiO_2); occurring as crystalline, cryptocrystalline, microcrystalline, and impure forms

silicify: turn to silica

siltstone: sedimentary rock formed from silt with particles smaller than sand grains and larger than clay particles

slab: a petrified wood specimen generally cut perpendicular to the long axis in cross-section [transverse] with a substantially greater diameter than thickness

stratigraphy: geology that deals with the origin, composition, distribution, and succession of strata

stratum: a layer of sedimentary rock or earth of one kind lying between beds of other kinds

tangential section: a cut made lengthwise through a stem, branch, or trunk perpendicular to the radius

Index

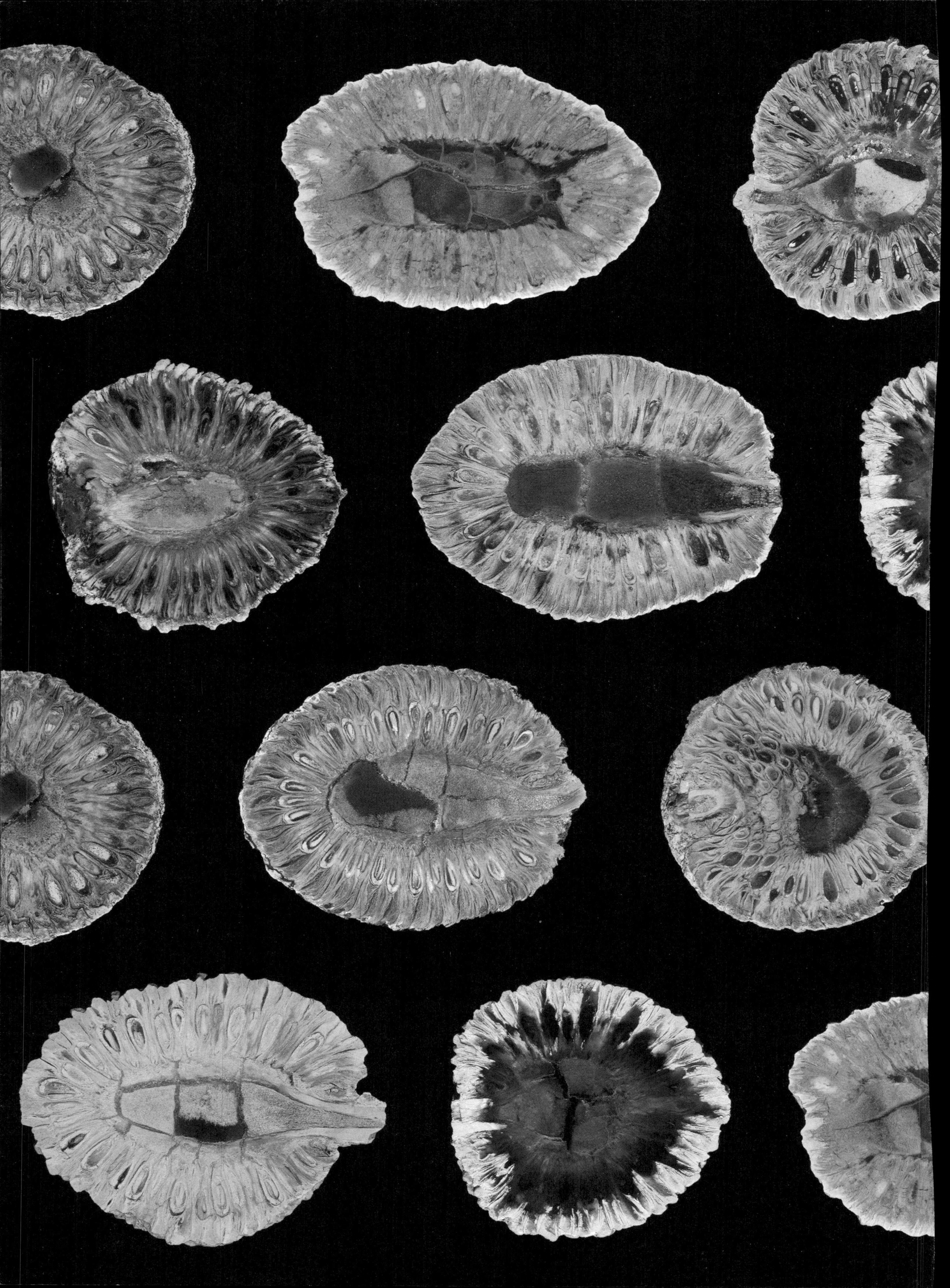